Water, Governance, and Crime Issues

Katja Eman • Gorazd Meško
Lorenzo Segato • Massimo Migliorini
Editors

Water, Governance, and Crime Issues

Springer

Editors
Katja Eman
Faculty of Criminal Justice and Security
University of Maribor
Ljubljana, Slovenia

Gorazd Meško
Faculty of Criminal Justice and Security
University of Maribor
Ljubljana, Slovenia

Lorenzo Segato
Research and Action - REACT
Padova, Italy

Massimo Migliorini
LINKS Foundation
Torino, Italy

ISBN 978-3-030-44800-4 ISBN 978-3-030-44798-4 (eBook)
https://doi.org/10.1007/978-3-030-44798-4

This Springer imprint is published by the registered company Springer Nature Switzerland AG
The registered company address is: Gewerbestrasse 11, 6330 Cham, Switzerland

Foreword

Discussion about Water, Governance and Crime Issues

Water is found almost everywhere on Earth. Approximately 70% of our planet's surface is covered by water in the form of oceans, seas, lakes, rivers and streams, canals and ponds, waterfalls and underground water caves, thermal springs, icy Arctic and Antarctic areas, as well as lesser-known underground reservoirs. Water is considered the second most important natural element for the survival of the human (and almost every other) species, preceded only by air. Furthermore, approximately 75% of the human body is composed of water. Therefore, water is a crucial element of our planet that cannot be bartered with. The fact that more than two-thirds of our planet's surface are covered by water leads to the conclusion that water often represents an ecosystem—a living environment—for numerous species, thus acting as the cradle of biodiversity, which is particularly important for the creation and preservation of ecosystems' sustainability. Indeed, every species contributes to the equilibrium of its ecosystem. Sometimes, it seems that all other living species, except humans, acknowledge and fulfil their role. Only human beings seem to forget where they come from and what is indispensable for their existence on the planet Earth. In this context, we are faced with an increasing problem of drinking water scarcity and the threat it poses to areas and countries that are naturally rich in water.

In the past 100 years, the consumption of drinking water has increased sixfold. The main reasons arise from population growth and the loss of water resources due to climate change. Furthermore, the volume of toxic emissions released into the water has been increasing and making many water sources useless. Karnani (2014) emphasises that the world is already facing a water crisis because data reveal that today, one in seven persons does not have adequate access to safe drinking water. Thus, drinking water represents the second major problem of this millennium after global warming.

Nowadays, we are witnessing two extremes: on the one hand, there is the lack of water, the consequences of which are reflected in poverty, starvation and death. On the other hand, we are observing unforeseeably large amounts of water (flooding, tsunamis, etc.), which cause pollution of drinking water sources and consequently give rise to various diseases and death. Brisman and South (2012) note that climate change caused by global warming and human interference with the environment is the reason for such extremes. Approximately four billion people (two-thirds of the world's population) struggle due to severe water shortages for at least one month every year. The most affected countries include China, India, Bangladesh, Pakistan, Nigeria, Mexico and parts of the USA. In other words, the "levels of groundwater are falling, lakes are drying up, less water is flowing in rivers, and water supplies for industry and farmers are threatened" (Gonzalez Rivas, 2014). This poses a tremendous threat to the availability of freshwater at the global level. Moreover, access to clean water is an additional problem faced by some countries across the world. Clean water is one of the basic requirements for health and development, although according to the World Health Organization, only half (54%) of the world's population can access water through a household connection to a piped system. The scarcity of drinking water and the difficulties related to access to water are merely two of the various reasons why people are leaving their homes (i.e. migration due to environmental issues), although the most terrible reason for moving stems from the prohibited access to always-available water sources that were sold to a corporation.

Water is an extremely unevenly supplied public good, where major differences occur between rural and urban areas. However, many goods provided as a public utility, such as clean water, are not public goods in the strict sense of the term. Clean water, for example, is subject both to rivalry and excludability. Since the quantities of clean water are limited, one individual's use of that water reduces its availability to others. Given the equipment and infrastructure required to transport clean water across distances and deliver it to households, it is certainly possible for governments to exclude certain people from accessing water or to privilege others. For example, the study of water access in the Mexico City Metropolitan Area revealed significant variations between societal groups, which can be attributed to differences in the political power of those groups (Gonzalez Rivas, 2014). Similar cases include unequal distribution of freshwater in Brasilia, where indigenous people were forced to leave the land on which their cattle has been drinking water daily for decades before a private company bought the land, or the almost dried out Aral and Chad lakes, as well as the Ogallala Aquifer in the USA. The severity of such violations of environmental (i.e. water) protection legislation is on the rise, while similar unlawful acts are also observed in Europe (e.g. Greece, France, Romania, etc.).

The fact that the human race is dependent on water is undeniable. Not only do we drink it and need it for our survival, but we also use it in energy production, industrial production and farming. Nowadays, although we live on a "blue planet", the amount of freshwater is decreasing and its preservation is therefore ever more important. Furthermore, any form of pollution or theft or other illegal activity against water resources is thereby a much more serious type of crime. Any violation of water protection legislation is defined as a crime against water. Against this background, the present book aims to present different forms and cases of violations, or "crimes", against water resources.

While the awareness of environmental crimes is gaining momentum at the international level, crimes against water are an emerging issue. Such crimes have significant impacts, although intelligence is scarce and information fragmented.

Understanding what types of violations may be committed, how they are perpetrated and what impacts they may have on water resources, as well as delineating the crime-trends foreseen in the near future and the factors contributing to the risk of such violations occurring, is a fundamental step to deal with the challenge of protecting water in the coming years. This first step has been taken by discussing water, governance and related crime issues in this book.

The idea for this book emerged from the international research project entitled Water Crimes funded with support from the European Commission (Directorate-General Home Affairs—HOME/2014/ISFP/AG/EFCE/7241). Water Crimes was a research project aimed at providing an inventory of water-related crimes in Europe, the first report on water crimes in Europe and the first strategic analysis in Europe on crimes against water resources. The organisations working on the project originate from four countries: Italy, Hungary, Spain and Slovenia. SiTI—Istituto Superiore sui Sistemi Territoriali per l'Innovazione (Italy) was the partner responsible for coordinating the Water Crimes Project. Other partners included RiSSC—Centro Ricerche e Studi su Sicurezza e Criminalità (Italy), REC—Regional Environmental Centre for Central and Eastern Europe (Hungary), UDC—University of Coruña (Spain) and FJCS-UM—Faculty of Criminal Justice and Security of the University of Maribor (Slovenia). More information about the project is available at http://www.watercrimes.eu/. The authors from Cyprus, Canary Islands (Spain), South Africa, Australia, UK, the USA and other Slovenian universities joined us later.

The present book is divided into two major parts: The first part presents theoretical perspectives on crimes against water, known also as water crimes, while the second part focuses on geographic and practical perspectives on water crimes.

The chapter authored by *Rob White* and *Katja Eman* introduces a green criminology perspective on water crimes as a global issue. The authors note that water is worth more than all gold stocks and oil prices, as it represents the source of (human) life. The existence of all living creatures is simply impossible without water. However, the fact that freshwater reserves are limited gives rise to serious problems. Having this in mind, the authors try to highlight the importance of freshwater preservation. They analyse and compare different types of water crimes. In conclusion, the authors attempt to provide an answer to the question as to how to successfully protect freshwater resources.

Avi Brisman, Bill McClanahan, Nigel South and *Reece Walters* discuss the politics of water rights, emphasising the issues of scarcity, sovereignty and security. The chapter thus explores the ways in which corporate power, supported and sponsored by government initiatives and legal frameworks, monopolises an essential global resource with devastating environmental and human consequences. The authors emphasise that the unsustainable and exploitative use of scarce global resources of freshwater continues to create conflict and human dislocation on a grand scale. Instead of witnessing nation-states adopting more equitable and efficient conservation strategies, powerful corporations are permitted to privatise and monopolise diminishing water reservoirs based on flawed neoliberal assumptions and market models based on "global goods". The commodification of water has enabled corporate monopolies and corrupt states to exploit a fundamental human right while, in the process, creating new forms of criminality.

Lorenzo Segato, Walter Mattioli and *Nicola Capello* try to define the position of water crimes in the environmental crime structure. Unlike any other crime, environmental crimes are aggravated through their additional cost and impact on the environment, as well as their cost to future generations. In this framework, crimes against water have been considered as an emerging global issue, while "water crimes" have only been deemed as an emerging environmental crime recently. The authors emphasise that water does not have a unique position and cannot be isolated from environmental crimes: it may be the environmental resource damaged by a crime (e.g. surface water pollution or fraudulent water quality reporting); the object of a crime (e.g. drinking water theft) or the means of the crime (e.g. intentional flooding or deliberate poisoning of a water supply). The chapter reveals that the analysis of "water crimes" is hampered because offences against water are often recorded under other crimes, such as fraud, corruption, trafficking, etc. The main cause of these issues stems from the absence of a common definition of "water crimes", which consequently leads to the absence of systematic data collection. The authors conclude that giving relevance to water crimes in the scope of environmental crimes is an important turning point that led to a systematic collection of information useful for the criminological analysis of the phenomenon and for assessing, detecting and prosecuting such crimes.

Katja Eman and *Rob White* tackle the issue of water as a new high-profit low-risk activity of organised crime. Water is one of the last "profitable niches", where organised crime groups count on large profits, particularly due to its scarcity, and transform it into a tradeable commodity. The re-conceptualisation of water as a tradeable commodity opened a new way of making profit not only for legal but also for illegal stakeholders. Although organised crime groups have been involved in environmental crime for many years, these activities remain relatively unexplored. In order to tackle organised environmental crime effectively, the second part of this chapter describes several methods and modes, such as environmental enforcement sweeps, National Environmental Security Task Force (NEST), the "method of disruption" and harmed-focused policing, which were already proven to be successful in responding to (organised) environmental crime. Finally, the authors discuss the usefulness of such methods in the water sector.

Walter Mattioli, Nicola Capello and *Lorenzo Segato* note that different data, different formats, different classifications and different assessment processes lead to the lack of interoperability among police forces and to the recording of water crimes under other offences. Moreover, without an adequate inventory, there cannot be any understanding of the number and impact of offences against water. With the aim of solving this problem, approaches to crime classification systems were analysed to understand the capacity for identifying and collecting information on water crimes and, eventually, understand how to overcome the existing limits. Furthermore, the International Classification of Crime for Statistical Purposes (ICCS) was used and tested. The results of the test revealed that the information about "water" is lost in the data collection process and that the information about the type of water, the service stage involved or the intended use are not accounted into the ICCS system. Finally, the authors divide water crime cases into seven main categories (threats): 1. Water Corruption; 2. Water Organised Crime; 3. Water Pollution; 4. Water Theft; 5. Water Fraud; 6. Water Terrorism and 7. Water Cyberattacks.

In the last chapter of the first part of the book, *Gorazd Meško* and *Katja Eman* focus on policing water crimes. Water crimes include various types of crimes ranging from the pilfering of water from pipelines and water pollution to fraud and illegal trafficking of water, etc. Therefore, they sometimes represent a real challenge for police investigators. Undoubtedly, water crimes are a type of crime that is difficult to detect, investigate, prosecute and study. It is thus necessary for law enforcement officers to have the necessary knowledge related to water crimes, which should include the knowledge of natural and social sciences, as well as a grasp of other disciplines, such as biology and chemistry. Furthermore, they must have a well-organised coordination and cooperation with other formal social control entities, such as inspectorates, institutes, etc. The authors suggest that policing water crimes should go beyond the classic methods and also include specialised forms of policing, such as problem-oriented policing, community-oriented policing and, where necessary, even intelligence-led policing.

The second part of the book is a snapshot of the evolution of water crimes in different contexts and countries. Based on the scarce data available, the authors analyse water offences in their respective countries and try to devise recurring patterns of criminal behaviours.

The chapter authored by *Luis Javier Capote Pérez* discusses the legal protection of water in the Canary Islands. The author notes that in a world populated by an ever-growing number of people and characterised by a proportionally increasing preoccupation about the demand for natural resources, access to water is being conceived as a fundamental right, related to other similarly important rights, such as health, quality of life or the previously mentioned environment. From the legal perspective, Water Law is now a special discipline of legal science, which combines private and public aspects. Civil Law, Criminal Law and Administrative Law are combined in the regulation of water, which is considered both as a resource and as a product. The chapter provides a brief overview of water regulation in the Canary Islands, a Spanish archipelago where this resource is deeply linked to its geographic configuration and its geological nature and is consequently reflected in its history and its specific legal management.

Ioanna Hadjiyianni, Andreas Kapardis and *Nicos Pavlides* studied water crimes in Cyprus at different levels of governance, from the international to the European and local levels. The authors identified the progress made at the EU level in improving enforcement through the criminalisation of infringements of the key environmental law directives, including the Water Framework Directive. The observed progress, however, does not necessarily translate into effective enforcement through criminal law at the national level, given the discretion granted to the Member States in determining the severity of offences and ensuring their enforcement. Finally, the authors emphasise the institutional weaknesses of authorities responsible for enforcement and suggest the necessary improvements.

Iztok Rakar, Bojan Tičar and *Tina Sever* present legal water protection in Slovenia. Since Slovenia boasts rich water resources, their legal regulation, protection and governance are important and (unfortunately) also complex. This chapter presents selected constitutional and administrative law perspectives on water protection in Slovenia. This is a very important issue, as Slovenia is one of the few countries in the world, which regulates the right to drinking water at the constitutional level. The legislation analysis shows that part of the water protection regulation lacks clarity and the new sector-specific legislation has yet to be adopted and harmonised. The authors emphasise that legislation implementation, and particularly its supervision, is faced with challenges related to human resources management and regulatory activities. They conclude this chapter by discussing possible solutions.

Saša Kuhar and *Gorazd Meško* present the water crimes situation in the Republic of Slovenia. Slovenia is one of the smallest countries in Europe, but one of the richest in terms of its water resources. Unfortunately, they are not uniformly distributed across the entire territory. A study of statistics, reports and documentation revealed that water crimes are not very frequent in Slovenia. The water pollution offences accounted for 85 per cent of all cases. In most cases, such offences targeted drinking water and sources of freely available natural water, such as rivers, lakes and ponds. Potential dangers for the water sector included the pollution of groundwater, pollution of streams and rivers by factories and theft of water from the pipeline system. In most water pollution cases, the perpetrators remained unknown. According to police statistics, most suspects of water-related crimes were male (82%), while female suspects and legal persons accounted for 8 and 10% of all recorded water, respectively. Suspects were between 34 and 44 years of age. The most significant issue stems from the dark figure of crime against water, since many cases remain undetected and unreported.

Luz María Puente Aba and *Eva María Souto García* present water crimes in Spain and discuss the protection of water from a criminal perspective. The results of their study reveal that two issues are particularly relevant: the frequent cases of negligence and the involvement of public authorities in the commission of water crimes. The latter is related to corruption, a crime which can also affect the water sector. Apart from water pollution, cases of water poisoning are also relevant within the realm of environmental crimes. The authors note that water accessibility and availability can also be affected not only in terms of human health but also from an economic perspective, as fraudulent behaviour or theft during the process of water distribution can endanger universal access to water. Finally, it is relevant to analyse how water can be used to commit other types of crimes, which can affect human life, such as terrorist attacks, whereby water is manipulated to spread a terrorist threat, regardless of its realistic or fictional nature.

A chapter authored by *Anthony Minnaar* presents the case of water pollution and contamination from acid mine drainage in the Gauteng Province in South Africa. The contaminated water that flows out of mines, termed Acid Mine Drainage (AMD), is dangerous for people's health, as well as for plants, animals and aquatic life. By the mid-2000s, Acid Mine Drainage had reached a crisis point in the Gauteng Province of South Africa. This crisis occurred simply because some mining companies had abandoned mines without rehabilitating them underground nor neutralising the above-ground mine dumps and slime dams, which subsequently caused the acid mine water to flow into streams, dams and groundwater sources across the province. Despite government efforts to persuade mining companies to take responsibility for "cleaning up" mine water seepage, mine dumps and slime dams, outflows of acid mine water continue to pollute water systems throughout Gauteng. The author examines the tardy efforts of the central government to enforce the mine rehabilitation and acid mine water neutralisation regulations on mining houses, which have in turn conveniently ignored them for many years in pursuit of maximising their profits.

At last, *Katarina Zajc* discusses water privatisation issues based on past examples from the European Union Member States. Firstly, the chapter focuses on empirical evidence, particularly the rule of law, where special emphasis is placed on water crimes and the efficiency of private sector privatisation options (concession agreements or outright sales of water utilities) in transition economies. Secondly, a one-sided prisoner dilemma model, developed and described in the scope of a publication entitled "Privatization and Water Infrastructure" (Zajc, 2004), was used, which predicted that in economies with a high degree of the rule of law, *ceteris paribus*, privatisation options do not matter, since both are efficient. When the rule of law is absent or diminished, outright sales of water utilities are more efficient. Finally, the author discusses the collection of the 15 years' worth of empirical evidence on the rule of law and privatisation options in water utilities and tests the proposed hypothesis set in 2004.

Water crime issues represent a relatively new and still unknown (i.e. unresearched) field of criminology, law, criminal investigation and other sciences. Findings of the Water Crimes Project and data from the authors' chapters reveal that we have been tackling the growing issues faced by the studied countries. Nevertheless, they require more in-depth studies and proper training of professionals, particularly police officers, prosecutors and judges. Water crimes can have significant impacts on people's lives (including on their survival). Thus, understanding which violations can be committed and how, in what forms they can be perpetrated and what impacts they may have on water resources, is a fundamental step in dealing with the challenge of protecting water in the coming years. We also hope that the message of this book will be loud enough to reach those responsible for the governance of water resources and water protection. We believe it is extremely important to refine knowledge-generation tools, promote broad and systematic cooperation with national and international law enforcement agencies and push for the establishment of a standard approach to building and diffusing knowledge applied for assessing the risks of crimes against water.

This publication is not merely a collection of country reports but also an essential reading about water crimes and related issues. A transfer of ideas, research results, practice, policies and legislation can contribute to the understanding of water crime consequences (and victims) and the development of water resource protection tools.

As mentioned at the beginning of this preface, this book is merely a small part of the water crime mosaic; however, we hope to have succeeded in raising the main topics for further discussion and development in this field of research.

The (green) criminological research presented in this book is mainly based on the authors' ambition and dedication to develop responses to water crimes and to promote water protection policies and practices in their respective countries. It is the editors' and authors' belief that water crime issues deserve further scientific study, either as a part of the green criminology or in the scope of other scientific disciplines. We appreciate the effort of all those who contributed to this publication to inform readers about water regulation, water governance and water crime issues in their respective countries.

Ljubljana, Slovenia — Katja Eman
Ljubljana, Slovenia — Gorazd Meško
Padova, Italy — Lorenzo Segato
Torino, Italy — Massimo Migliorini
February 2020

References

Brisman, A., & South, N. (2012). A green-cultural criminology: An exploratory outline. *Crime, Media, Culture, 9*, 115–135.

Gonzalez Rivas, M. (2014). Ethnolinguistic divisions and access to clean water in Mexico. *Latin American Research Review, 2*, 129–151.

Karnani, A. (2014). Corporate social responsibility does not avert the tragedy of the commons. Case study: Coca-Cola India. *Economics, Management, and Financial Markets, 9*, 11–33.

Acknowledgements

This book is a result of almost 3 years of research work in the framework of the Water Crimes international research project, funded with support from the European Commission (HOME/2014/ISFP/AG/EFCE/7241). We are grateful to the researchers who have worked on the project. They come from the following organisations: Leading Innovation & Knowledge for Society (Italy), Research and Action (Italy), The Regional Environmental Center for Central and Eastern Europe (Hungary), University of A Coruña (Spain) and Faculty of Criminal Justice and Security of the University of Maribor (Slovenia). Later joined us and greatly contributed to the book are the authors from the University of Cyprus (Cyprus), Regulatory Compliance Services Ltd Nicosia (Cyprus), Universidad de La Laguna (Canary Islands, Spain), Eastern Kentucky University (USA), Queensland University of Technology (Australia), University of Essex (UK), Deakin University (Australia), University of Tasmania (Australia), NAP University of South Africa (South Africa) and University of Ljubljana (Slovenia).

We would like to express our gratitude to everyone who has actively contributed to the successful publication of this book. We would first like to thank the authors of the papers included in this book who have worked for more than two years to complete their manuscripts. Their zeal and dedication were observed throughout the editing process from the first peer reviews and paper revisions to the final copyediting. The authors' work has greatly contributed to the recognition of water crime issues and challenges in their respective countries and worldwide.

We want to thank the international organisations that have expressed a strong interest in our work on water crimes: UNODC, INTERPOL, the World Water Institute, the World Bank and the IMPEL Network. Our gratitude is also extended to peer reviewers who helped us improve the drafts. An equally valuable contribution was made by the technical editing staff who prepared, proofread, laid out and typeset this book. We are indebted to Ms Katherine Chabalko for encouraging us to publish this volume. Special thanks go to Ms Sofia Geck and Ms Judith Newlin for their support in the editing process. Last but not least, we are hugely appreciative and thankful for the support provided by Springer's copyediting service.

We are grateful to the authors for their original contributions and persistence during the editing process. A core group of highly committed and genuinely enthusiastic researchers from different parts of the world was developed, persevered to the present time and is looking forward to new research challenges. Due to their active engagement, this publication is not merely another book. It is a significant milestone in terms of recognising water crime(s) as a set of "new" growing issues of contemporary society, the need for further exploring and analysing these issues and the role of green criminology in responding to water crime challenges at the local, regional and global level.

Ljubljana, Slovenia Katja Eman
Ljubljana, Slovenia Gorazd Meško
Padova, Italy Lorenzo Segato
Torino, Italy Massimo Migliorini

Contents

About the Authors

Avi Brisman, MFA, JD, Ph.D., is an Associate Professor in the School of Justice Studies at Eastern Kentucky University, an Adjunct Associate Professor in the School of Justice at Queensland University of Technology and a Conjoint Associate Professor in the Newcastle Law School at the University of Newcastle. He is also Editor-in-Chief of Critical Criminology: An International Journal.

Nicola Capello holds a degree in law from the Università Statale in Milan, where he also obtained a diploma from the "Scuola di Specializzazione per le Professioni Legali". He has completed an advanced master on the analysis, prevention and contrast of organised crime and corruption, at the Faculty of Political Science of the Università Statale in Pisa. He is a researcher for REACT—Research and Action Against Crime and Corruption. His main fields of interest are the study and analysis of corruption within institutional frameworks.

Luis Javier Capote-Pérez, Ph.D., is a Lecturer of Civil Law at the Basic Legal Disciplines Department, Universidad de La Laguna, Spain. He has been Visiting Lecturer at the Faculty of Law of the University of Sassari, Italy. He is member of the UNESCO Cultural Chair "Cultural Property Law" of the University of Opole, Poland. He did his doctoral thesis about Tourism Law and published a book entitled *El tiempo compartido en España: un análisis de la fórmula club-trustee desde la perspectiva del Derecho Español* (Capote-Pérez; Tirant Lo Blanch, 2009). He is the co-author of a book about *Intellectual Property Law entitled Introduction to Digital Copyright Law* (Mezei, Hajdu & Capote-Pérez; Iurisperitus Kiadó, 2018). He is now working in a project on Cultural Property Law and cultural heritage protection.

Katja Eman, Ph.D., is an Associate Professor of Criminology at the Faculty of Criminal Justice and Security, University of Maribor, Slovenia. Her doctoral thesis was entitled *Crimes Against the Environment: Comparative Criminology and Criminal Justice Perspectives* (2012), and she is the author of *Environmental Crime and Criminology: Crime Phenomena and Development of a Green Criminology in Slovenia* (Scholar's Press, 2014). She participated in a research on water crimes in

Europe (2016–2017) and is currently a lead researcher in the EU research project SHINE on sexual harassment in nightlife city areas (2020–2022).

Ioanna Hadjiyianni, Ph.D., is a Lecturer in Law at the Department of Law, University of Cyprus. Prior to that, she was a Max Weber Postdoctoral Fellow at the European University Institute. She has recently published a monograph on The EU as a Global Regulator for Environmental Protection: A Legitimacy Perspective (Hart Publishing, 2019) based on her PhD thesis which was completed at King's College London. She is currently leading a research project on the Global Reach of EU Law: The Case of Ship Recycling funded by the University of Cyprus (2020–2021).

Andreas Kapardis, Ph.D., holds a Ph.D. in Criminology from Cambridge University and is Professor Emeritus (Active) Law Department, University of Cyprus. For a number of years, he taught in the School of Law at La Trobe University in Australia and was a Visiting Professor at the Institute of Criminology, Cambridge University, 1999–2018. His research and teaching interests lie mainly in criminology, criminal justice, legal psychology and penology. He has gained numerous European and local research grants on a competitive basis, has researched both a broad range of offenders and crimes as well as judicial decision-making and policing He has published extensively internationally. His books and co-authored/co-edited books include: *Greeks in Australia* (1988); *They Wrought Mayhem: An Insight to Mass Murder* (1989); *Economic Crimes in Cyprus* (2001); *Society, Crime and Criminal Justice in Cyprus the First Years of British Rule* (2001); *Sentencing in Cyprus* (2003); *Risk Assessment for Juvenile Violent Offending* (2013), Routledge, with Anna Baldry; *Psychology and Law* (2014), Cambridge University Press (4th ed.) and *Psychology, Crime, Policing and the Courts* (2016), Routledge, with David Farrington; Kapardis, A. Krispi-Samara, A. (eds) (2018). Protecting the Environment for Human Intervention. Athens: A.N. Sakkoulas; Fox, B., Farrington, D.P., Kapardis, A. and Hambly, O.C. (eds.), Evidence Based Offender Profiling. London: Routledge (forthcoming 2020) and Kapardis, A. and Stephanou, E., Cypriot Penology and Criminal Justice. Athens: Nomiki Vivliothiki Publishers (forthcoming 2020).

Saša Kuhar, Ph.D., is a Senior Advisor at the Emergency Preparedness Division, Slovenian Nuclear Safety Administration and a lecturer in Security Studies at the Faculty of Criminal Justice and Security, University of Maribor, Slovenia. Her main research focus are nuclear safety, water crimes, art crimes and critical infrastructure. She is also an expert lecturer at the International Atomic Energy Agency. She was a Research assistant at the Water Crimes Project in the years 2016 and 2017. The project was co-funded by the Internal Security Fund of the European Union. She has about 20 publications on environmental crime, criminal investigation, criminal justice and nuclear safety.

Walter Mattioli, Ph.D., is a forester with a degree in Forestry and Environmental Sciences and with a Ph.D. in Environment and Territory. Graduated in 2003, he has more than 50 scientific and technical publications, of which several have been published in ISI journals with impact factor and in peer-reviewed scientific journals. Currently, he has a Post-Doc Position at University of Tuscia (VT) after has been a researcher at CREA, Research Centre for Forestry and Wood. His main fields of interest are remote sensing, geomatics, forest management, biodiversity conservation and, recently, illegal activities in forestry sector.

Bill McClanahan, Ph.D., is Assistant Professor in the School of Justice Studies at Eastern Kentucky University. His research on climate change, water, policing and photography has been published in several journals and edited volumes.

Gorazd Meško, Ph.D., is a Professor of Criminology at the Faculty of Criminal Justice and Security, University of Maribor, Slovenia. He was a lead editor of books entitled *Trust and Legitimacy in Criminal Justice: European Perspectives* (Meško & Tankebe; Springer, 2015), *Handbook on Policing in Central and Eastern Europe* (Meško, Fields, Lobnikar & Sotlar, 2013) and *Managing and Understanding Threats to the Environment* (Meško, Dimitrijević & Fields; Springer, 2011). He participated in a research project on water crimes in Europe (2016–2017) and is currently a lead researcher in a national research project on safety and security in local communities (2019–2024).

Massimo Migliorini is Senior Researcher, has a degree in chemical engineering (2003). Since 2005, he has worked at SiTI's Security and Safety department, focusing his activities on risk assessment and impact assessment methodologies for strategic infrastructures (including cross-border contexts and cultural heritage). He has developed a profound knowledge and expertise in the field of risk assessment and security assessment, in particular against crimes and man-made attacks. He has specific knowledge of complex and networked systems, including cascading effects. Critical infrastructures are another domain of expertise, with operations in the energy and water domain. He has played a central role in several projects funded by the EU (7FP, Alcotra, DG Home), such as SIAM, PICRIT, FORTRESS, PRODIGE and WATERCRIME. He actively works in two priority research subjects for SiTI: security of cultural heritage (museums, archaeological sites, artworks) against anthropogenic threats (including terrorism) and the use of the INSPIRE Directive for civil protection purposes at the transboundary level. In 2018, he coordinated the European Project RESCULT devoted to enhance the capability of Civil Protection to prevent and mitigate impacts of disasters on sites of Cultural Heritage, through the realisation of an integrated European Interoperable Database (EID). From 2018, he is coordinating the Data for Resilience Working Group of the European Scientific and Technology Advisory Group, set up by the European Commission (DG-JRC) and the United Nations (UNISDR) to promote the application of Sendai Framework for Disaster Risk Reduction Principles.

Anthony Minnaar, Ph.D., from January 2009 to June 2014 was the Programme Head: Security Management and Postgraduate Coordinator in the Department of Criminology and Security Science in the School of Criminal Justice of the College of Law at the University of South Africa in Pretoria. He was appointed a full Professor (Criminal Justice Studies) in July 2003. In January 2013, he was appointed as a Research Professor at UNISA. A National Research Foundation (NRF) rated Researcher (2003–2007 (C3); 2008–2012 (C2)) and in 2012 received an NRF re-rating of C1 (2013–2018). He is a recipient in 2006 and 2017 of the University of South Africa (UNISA) Chancellor's Prize for Research Excellence (College of Law). He has published widely and presented at numerous international conferences on a wide variety of topics and issues in criminal justice, criminology and policing. In the early 1990s, he published largely on issues of political violence and conflict inter alia migrant labour hostel violence (Goldstone Commission Report); warlordism; massacres and the proliferation of firearms and self-defence units. In the mid-1990s, his research interests turned to other forms of violence such as land disputes; informal settlements; illegal squatting and evictions; minibus taxi industry conflicts; violence around witchcraft accusations; xenophobia and undocumented migrants and vigilantism. More recently, he has researched border controls; migrants'/refugees' rights; motor vehicle hijackings; use of force by police; the murder of police officers; use of informers and witness protection programmes; the struggle to legislate for stricter gun controls and the declarations of persons to be unfit to possess a firearm and security measures at ports-of-entry. His current research interests are in the broad field of applied criminal justice dealing with the specific issues of corruption prevention, border controls and undocumented migrants; xenophobia and refugees; use of firearms in violent crime; use of force by police/security officers and issues of militarisation vs demilitarisation of law enforcement; civilian oversight of public and private policing; private security industry issues (specifically crime prevention and private policing; community policing and community safety initiatives); regulating and monitoring the Private Security Industry in South Africa; security measures at sea/air ports-of-entry; cyber-attacks, cybersecurity, information protection and cybercrime; ransomware; privacy issues and cyber-surveillance; the use of CCTV surveillance systems for security and neighbourhood safety/crime prevention and issues of rural crime in South Africa. He was Editor-in-Chief (2012–2016) and Co-Editor-in-Chief (2017–2018) of the journal: Acta Criminologica: Southern African Journal of Criminology. He retired from UNISA in December 2018. Currently, he is a Research Associate in the Department of Criminology and Criminal Justice at the University of Limpopo, Sovenga, South Africa.

Luz María Puente Aba, Ph.D., is Associate Professor of Criminal Law at the University of A Coruña (Spain), where she teaches and researches in the fields of Criminal Law and Criminology. Some of her research interests are Environmental Criminal Law, Business Criminal Law and Corruption; she has participated, as member of a research team and as main researcher, in different research projects (at national, international or European level) regarding these topics. She has published books, journal articles and book chapters and also given lectures, in Spain and several other countries (Italy, France and Chile). She has been member of the European Society of Criminology and the International Association of Criminal Law and now is focused on the Spanish Group on Research Policy; besides, she was appointed to be part of a database of independent experts to assist the European Commission in the field of freedom, security and justice. She also worked for some time as a Public Prosecutor.

Iztok Rakar, Ph.D., is an Assistant Professor of Public Administration. He took his PhD from the Faculty of Law, University of Ljubljana. He researches in the field of administrative law, local self-government and comparative public administration. Since 2013, he holds Chair of Administrative-Legal Area at the Faculty of Public Administration. He is a titular member of the Group of Independent Experts on the European Charter of Local Self-Government by the Council of Europe for Slovenia (since 2013) and coordinates Rapid Response Expert Network to assist public administration reform in Ukraine within the EU4PAR project (since 2018).

Lorenzo Segato, Ph.D., holds a PhD in Criminology and is CEO of REACT—an Italian research centre on crime and corruption. His main fields of interest are corruption, organised crime and environmental crimes, including water crimes. He has a long experience in research on criminal phenomena, impact assessment, evaluation of public policies and working for national and international organisations. He has experience in coordination of international research projects, including the EU funded Water Crimes project. His articles, book chapter and reports have appeared in international and peer-reviewed publications, and he has been speaker at many national and international conferences.

Tina Sever, Ph.D., is an Assistant Professor at the Faculty of Administration, University of Ljubljana (UL). In 2006, she obtained a BSc degree at the Faculty of Law, UL and in 2014, a PhD at the European Faculty of Law. In 2007 and 2008, she was a trainee at the Administrative Unit of Ljubljana, European Commission (Directorate-General for Translation in Luxembourg) and European Ombudsman. She passed a Bar Examination Traineeship at the Higher Court of Ljubljana in 2009. In October 2008, she became a teaching assistant at the UL and in December 2016 she was appointed Assistant Professor. Her research interests are public administration and administrative law. She participates in different projects and is (co)author of several scientific and conference papers.

Nigel South, Ph.D., is Professor and Director, Centre for Criminology, University of Essex; Adjunct Professor, Crime and Justice Research Centre, Queensland University of Technology and a Fellow of the Academy of Social Sciences. In 2013, he received a Lifetime Achievement Award from the American Society of Criminology, Division on Critical Criminology.

Eva María Souto García, Ph.D., is Assistant Professor of Criminal Law at the University of A Coruña (Spain), where she teaches and conducts research in the fields of Criminal Law and Criminology. Most of her research interests are Business Criminal Law, Patrimonial offences, Fraud and Corruption; she has participated, as member of a research team in different research projects (at national, international or European level) regarding mainly these topics. She has published books (monographs), journal articles and book chapters in association with other researchers. Besides, she has given lectures, in Spain and several other countries (Portugal, Czech Republic and Ireland). She has done several research stays in Portugal, Italy and Germany. She has been member of the European Society of Criminology and she is nowadays member of ECRIM (Research Group on Criminology Legal Psychology and Criminal Justice in the twenty-first century).

Bojan Tičar, Ph.D., is doctor of legal sciences (Ph.D. from the Faculty of Law, University of Ljubljana). In 2005, he joined the University of Maribor, Faculty of Criminal Justice and Security where he teaches as professor of administrative law, legal regulation of management in the public sector. His interests cover the public-social and security systems, public and corporate governance as well as other legal issues in administrative-legal regulation. At present, he works on a couple of articles dealing with public sector legal regulation. His research and consultancy activities are concentrated in P.A.R. (Public Administration Reform), including mandatory training programs for civil service, the introduction of administrative inspection and reform of public servants' salary system.

Reece Walters, Ph.D., is Professor of Criminology in the School of Humanities and Social Sciences at Deakin University, Australia. He is also an Adjunct Professor at Queensland University of Technology, Australia; a Professorial Fellow at the Open University, UK and Adjunct Professor, Faculty of Arts, University of Tasmania, Australia.

Rob White, Ph.D., is Distinguished Professor of Criminology at the University of Tasmania, Australia. He has published widely in criminology and youth studies, especially in the areas of green criminology, juvenile justice and eco-justice. Among his recent publications are *Media and Crime* (with Katrina Clifford, Oxford University Press, 2017), *Climate Change Criminology* (Bristol University Press, 2018) and the two-volume edited collection *Environmental Crime* (Edgar Elgar, 2020).

Katarina Zajc, Ph.D., holds a Ph.D. in economics from George Mason University and an LL.M. in law from Yale Law School. She is a full professor of Law and Economics and teaches Economics and Law and Economics at the University of Ljubljana Law School. She is also a visiting professor at University of Donja Gorica in Montenegro and at CEVRO Institute in Prague, Czech Republic. She was a member of Judicial Council of the Republic of Slovenia for a period of 6-year term and is currently a member of the Administrative Council of the Republic of Slovenia.

Her interest are in corporate governance, judicial efficiency, privatisation and regulation. In 2004, she wrote a book titled *Privatization and Water Infrastructure*. Her articles are published in Journal of Comparative Economic, European Journal of Law and Economics, International Review of Law and Economics and Economic Systems. He is currently working on the efficiency of the regulation of gaming industry in EU.

Nicos Pavlides is a Cypriot lawyer, practicing law in and from Cyprus. He holds degrees in Economics and Law (Honours) from the University of Sydney and Master of Laws (LLM) from the University College of London. He is the recipient of the Chevening Scholarship (UK) and the Fulbright Scholarship (USA). He has worked for the prosecution of serious offences in Sydney and London. Since the accession of the Republic of Cyprus to the EU and the harmonisation of the Cypriot legislation with the European acquis, he has focused principally on financial markets law. An environmentalist, he offers services pro bono for green and environmental causes, keenly aware of the many environmental challenges faced in Cyprus and the Eastern Mediterranean region.

Part I
Criminological Perspectives of Water Crimes

Chapter 1
Green Criminology, Water Issues, Human Rights and Private Profit

Rob White and Katja Eman

1 Introduction

Fresh water is a basic essential of life. It is fundamental to the wellbeing of humans, to ecosystems and to flora and fauna. It constitutes the lifeblood of many industries, from tourism to mining, agriculture to aquaculture. Activities and events that threaten the quantity and quality of fresh water therefore have major negative consequences for health, business success and national security. What is more, water is "one of the world's greatest renewable resources and is a natural monopoly essential to health and wellbeing" (Whelan & White, 2005, p. 135). In other words, clean safe water presents as a biological, social, economic, and spiritual necessity for contemporary society (Archer, 1998, p. 30).

While essential to life, there are presently a number of threats to fresh water supply on a world scale. The scarcity of water is itself part of the problems. As this chapter demonstrates, water issues have a number of dimensions and from the point of view of human rights can be analysed in relation to notions of crime, criminality and social and ecological justice. Who owns and controls water, and how they do so, is of major concern.

R. White (✉)
School of Social Sciences, University of Tasmania, Hobart, TAS, Australia
e-mail: R.D.White@utas.edu.au

K. Eman
Faculty of Criminal Justice and Security, University of Maribor, Ljubljana, Slovenia
e-mail: katja.eman@fvv.uni-mb.si

K. Eman et al. (eds.), *Water, Governance, and Crime Issues*,
https://doi.org/10.1007/978-3-030-44798-4_1

2 Current State of Play

Today, we are witnessing two extremes. On the one hand, in many parts of the world there is a lack of fresh water, the consequences of which are reflected in impoverished communities, starvation and death. The over-exploitation of water resources is a major problem worldwide. An example of this is the Aral Sea which is 'disappearing' due to the irrigation used in agriculture which has greatly reduced the main river tributaries flowing into that sea (Karnani, 2014). Spoor (1998) stressed almost two decades ago that with no appropriate changes in how we manage them, there is the threat that water resources will run out by 2050. On the other hand, many existing water sources are subject to contamination. This may be due to industrial pollution of waterways or to climate-related events that periodically affect water supplies. Water quality is affected, for example, by floods and tsunamis that taint drinking water and subsequently foster disease. Brisman and South (2014) emphasize that global warming and human interference with the natural environment creates the conditions for drought and flood alike. Economic growth and technological development have reached a critical environmental limit, ultimately threatening the destruction of civilization as we currently know it (Meško & Eman, 2012).

Over past century water has been consumed at astonishing rates (Crank & Jacoby, 2015). In this period, for instance, the consumption of drinking water worldwide has increased six-fold. The main reasons for this are population growth. Simultaneously, there has been loss of water resources due to contamination and climate change (Eman & Meško, 2020). A scarcity of clean drinking water has become one of the crucial problems of the twenty-first century. As already emphasized, humans are dependent upon water. Yet, the amount of fresh water is decreasing, making its preservation even more important. As Karnani (2014) points out, that the world is already facing a water crisis, because daily one of seven persons does not have appropriate access to safe drinking water.

The harms and risks to water supply are associated with multiple causes that include industrial pollution and water theft, through to global warming. Its importance to personal, social and economic health and wellbeing has recently generated increasing criminological interest (White, 2003; Bricknell, 2010; Johnston, South, & Walters, 2015; Brisman, McClanahan, & South, 2016; Eman & Meško, 2013; Eman, Kuhar, & Meško, 2017, 2018). Especially in the light of environmental drivers such as land use, climate change and infrastructure, there is an urgent need to consider the nature, dynamics and impacts of water crimes. In order to do this, new typologies of crime have to be developed, new methodologies for research instigated, and new modes of social control devised if we are to adequately address present and future issues. White and Perrone (2015, p. 306) emphasize that questions pertaining specifically to environmental justice have been of longstanding interest and water is one of the crucial items in this discussion.

Studies on water have variously focused on water scarcity (Bennett & Lynch, 1990; White, 1998; González Rivas, 2014); water pollution (Pečar, 1981; White & Heckenberg, 2014; Barclay & Bartel, 2015; Eman, Kuhar & Meško, 2016), the privatisation of water (White, 2003; Whelan & White, 2005; Humar & Eman, 2017; Barclay & Bartel, 2015), and environmental justice issues (South & Brisman, 2013; Johnson, South & Walters, 2016). Discussions have also dealt with the issue of the quantity and quality of the water by relating these to the status of water as a tradeable commodity (Whelan & Willis, 2006; Whelan, 2009; Brisman et al., 2018), as well as non-economic consideration of the social and cultural value of water (Alston & Mason, 2008).

Declining water stocks, rising populations and climate variability collectively have created a situation of destabilisation over water resources all over the world (Alston & Mason, 2008, p. 214). Water use has become highly politicised. In regards to this, critics warn about the over-representation of business stakeholders in key policy decisions and the under-representation of other stakeholders, such as Indigenous people, local residents and those living in poverty, who have different social interests in how water is used (Alston & Mason, 2008; Johnson et al., 2016; White & Heckenberg, 2014; Whelan, 2009; Eman, Kuhar, & Meško, 2016; Brisman et al., 2016; White, 2016). Such situations lead to unequal access and use of water, and the undermining of environmental water protections for the population at large.

Water, especially in freshwater form, is a highly valued resource. In global terms, the topic of water crimes is steadily rising in importance as nation-states and enforcement agencies realise the scope and nature of the problem of water scarcity. For example, Interpol (2016) views water management as an emerging environmental crime that requires attention. In part this is because the amount of fresh water on earth (from both surface and groundwater sources) is limited and shrinking. It is estimated that around 85% of the world's population live in the world's driest regions and that by 2025, 3.5 billion of the earth's inhabitants will be living in water stressed conditions. In Africa alone, over 300 million people live in arid areas, which cover approximately 66% of the continent (Global Initiative, 2019).

As shown in Fig. 1.1, fresh water represents only a small part of all the water that is present on earth. Almost 70% of all the earth's surface is water, but only 2.5% of this water represents drinking water. The rest is made up of inland saltwater repositories and the oceans. Even now, only 1.2% of all types of drinking water is readily available as surface water, the rest presenting as groundwater and in glaciers and ice caps. In fact, there is only 0.007% of fresh water remaining and readily available for the world's population of 6.8 billion people (National Geographic, 2016).

Going from global to national and local, there are various emergent social and environmental harms associated with the production, consumption, management and preservation of fresh water (Whelan & White, 2005). All over the world countries are facing water issues although these issues differ from country to country. What matters most is the quality and quantity of the water that is available on a daily basis. Putting it differently, close attention needs to be paid to the kind of water people drink and the terms under which they access and drink it. It therefore becomes essential to analyse, understand and in many instances change the economic and political relationships that impact the interrelation of humans with water. For example, freshwater scarcity means that huge profits can be achieved when water resources are privatised, insofar as this involves supporting water industries

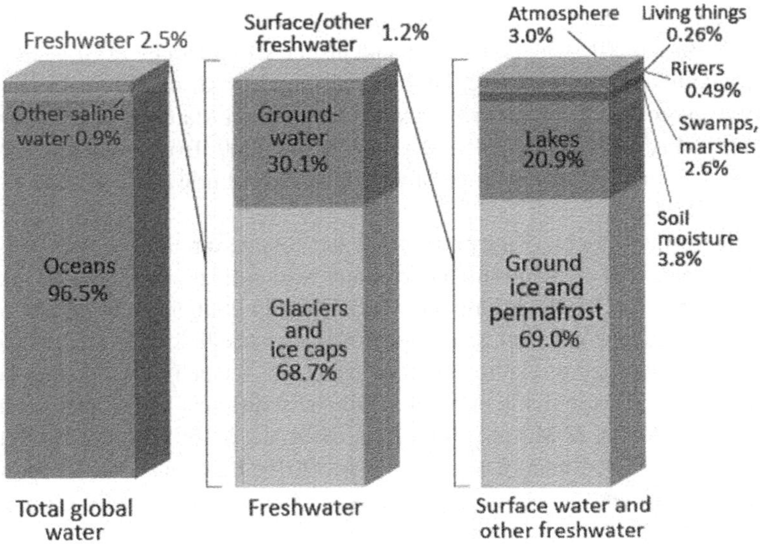

Fig. 1.1 Distribution of water on the planet Earth (source: U.S. Geological Survey n.d.)

that produce water only for those who can pay for it (Whelan & White, 2005). This type of allocation is not based on need and is clearly discriminatory and unjust.

As already mentioned, activities and events that threaten the quality and quantity of fresh water can have major consequences for health, social wellbeing, and personal and national security. However, far too often the crucial problem remains the conceptualisation and management of drinking water primarily or solely as an economic resource. This means that water is essentially seen as commodity and, as such, a source of making profit rather than as a basic requirement of human life (and other species, and ecosystems). A profit-based system also carries with it certain tendencies associated with profit maximisation. When water production is 'for profit' then it is the company bottom line that counts most, not the water itself or the interests of those who produce and consume it.

> The selling of water as a commodity means that it is first and foremost treated as a source of profit. Surplus value is created through the exploitation of workers who 'produce' the commodity. The rate of profit is contingent upon the producing company minimising expenses associated with fixed capital (e.g., equipment, plant infrastructure, technology) and variable capital (e.g., human labour) in commodity production, and maximising the realisation of surplus value through exchange of the commodity on the marketplace (e.g., consumer demand, presence or absence of competitors etc.). In real terms this has meant loss of jobs and increases of prices and this is what happened in the case of corporatisation of Sydney water (White, 1998, p. 216).

The chapter is premised on recognition that the present water crisis features a range of related harms and criminality. Exposing these has opened up a new field of study and in particular investigation of water issues by proponents of green criminology.

3 Lack of Water and Who Gets What and How

Alston and Mason (2008, p. 214) were right when they said that the twenty-first century heralds a new area of controversy over water. Access to clean fresh water is one of the basic requirements for health and development. However, water is today still not provided as a public good in most parts of the world. In addition, access to water is not evenly distributed. Only 54% of the world's population have access to water in their residence through a plumbing system. According to González Rivas (2014), this figure rises to 98% if we look only at developed countries but falls to 46% when considering only developing countries. It is also very unevenly distributed depending on the environment where the individual lives. González Rivas (2014) also notes that it in 2006 only 31% of the rural population in the world had access to water in their homes, compared with 78% of the urban population.

The World Health Organization (WHO 2015) emphasises that, as water is of great importance for the survival of each person, it should be free from discrimination and accessible to all, because this is a human right. This means that all citizens have the right to water, enough that is quantitatively sufficient and continuous for personal and domestic use, and it should be affordable and guaranteed by the state. It is a natural resource that ought to be available to all and accessible in the same proportion.

González Rivas (2014) stresses that around four billion people or two-thirds of the world's population, for at least one month yearly, have to struggle with severe water shortages. The most affected countries are China, India, Bangladesh, Pakistan, Nigeria, Mexico and parts of the USA. These problems reflect compounding issues such as falling groundwater levels, lakes drying up, less water flowing in rivers, and water supplies for industry and farmers being threatened (Rivas 2014). Together these trends present a huge threat to freshwater reserves on the global level. The scarcity of drinking water together with difficulty in access to it, are two important reasons why people are leaving their homes and migrating elsewhere. In some instances, the reason for moving is due to prohibited access to formerly available water resources. Bought by corporations this water is then subjected to restricted use based upon payment (Eman, 2016, p. 49). There are examples of this in Africa (South Africa, Ghana and Tanzania), India (Rajasthan), and Bolivia (Cochabamba and El Alto).

Water is also an unevenly provided public good where, for example, major differences occur between rural and urban areas. However, many goods provided as public services, such as clean water, are not public goods in the strict sense. Clean water, for example, is subject both to rivalry and excludability. Given limited quantities of clean water, one person's use of that water lowers its availability to another person. And given the equipment and infrastructure necessary to convey clean water across distances and deliver it into households, it is certainly possible for governments to exclude certain people from access or to privilege others. For example, study of water access in the Mexico City metropolitan area revealed large variations between societal groups, variation that is attributed to differences in the political power of those groups (González Rivas, 2014). Unequal distribution of freshwater is also evident in Brasilia where indigenous people were forced to leave the land where their cattle daily drank water for decades before a private company bought the land, and in the case of the almost dried out Aral Sea in Kazakhstan and Uzbekistan, Lake Chad in West Africa, the Ogallala Aquifer in the USA, and the Mexico City Aquifer (Tiha privatizacija vode, 2016).

The uneven distribution of water around the world has led to a situation in which water theft is on the rise. This includes the illegal acquisition of natural water courses as well as piped or harnessed water, both of which are designated as "non-revenue water", meaning water that is "lost" before reaching its intended consumer. Loss of water through theft can occur in a variety of ways including damaging or removing water meters and physically installing a connection to water distribution pipes as well as stealing aid tanks (Global Initiative, 2019).

Johnson, South and Walters (2016) believe that water should be a public good, the common heritage of people and nature as well as a fundamental human right. Moreover, it must be provided to each individual continuously and in sufficient quantity for personal use as well as household articles. According to the WHO (2015), a person requires between 50 and 100 litters of water daily to ensure the fulfilment of their basic needs. Water should be safe, which means that it must not contain micro-organisms, chemical substances and radiological hazards that could endanger the person's health. It must also be acceptable colour, odour and taste for each personal and household use. Everyone has the right to water and sanitation within or in close proximity to households, educational institutions, the workplace and health care institutions. According to data from the WHO (2015), the water source may be located more than 1000 m from home but collecting water should not exceed 30 min. Water and water services and facilities should be affordable to all. The United Nations Development Programme (UNDP) suggests that the cost of water should not exceed 3% of household income (United Nations Department for Economic and Social Affairs [UNDESA], 2014).

Ensuring the right to water, in addition to the economic and financial situation, geographical characteristics and quantity of available water resources in a country, to a large extent also depends on the way the water is managed. Over the past few decades due to global water scarcity and mismanagement of water resources, new ways to manage water began to appear. Foremost among these was the privatization of water management. This was often chosen in countries where the existing system of water management was not effective, due to poor supply, maintenance and water loss, or due to financial problems such as low income, poor financial management and lack of investment by the public sector. Accordingly, the reins of water management were handed over to private companies and organizations (UNDESA, 2014). We have more to say about this trend shortly.

The twenty-first century has seen the rise of an issue of international significance—the "global situation of destabilisation over water access" (Alston & Mason, 2008, p. 114). Eman & Meško, 2020 emphasize that the situation is so critical that in 2010 the United Nations adopted Resolution 64/292 that explicitly recognizes the human right to water and sanitation and acknowledges that clean drinking water and sanitation are essential to the realization of all human rights. This was an attempt to at least reduce the damage and friction related to access to water.

Despite UN Resolution 64/292 having been adopted, water is still a huge issue everywhere around the world, especially in situations where water is seen only as a tradeable commodity. For this reason, we find the 2015 UN-Water Annual International Zaragoza Conference an important development in the field of water protection, preservation, and sustainable development. The Conference brought together participants from United Nations Agencies and programs, experts and the academia, representatives of the business community, and governmental and non-governmental organisations. The theme of the conference was "water and sustainable development" (water, sanitation and hygiene, water resources management, water quality, and risks) and it resulted in the adoption of the Sustainable Development Goals (SDGs) that include the goal to ensure the availability and sustainable management of water and sanitation for all.

Since water is so vital to our survival, its protection is of great importance including through legal protections. At the global level, most countries do not have the legal right to water written into their Constitution. The right to water tends to be regulated by other laws and guidelines, which is a consequence of the higher volume of water available and lower interest in the economic market for water resources in the past (Eman & Meško, 2020). There are just a few countries that have further protected their water resources, not only as a human right but also from the predations of privatisation (that views it primarily as a source of private profit). Some examples include Italy, which strongly opposed the privatization of the water with a referendum; the Netherlands, which introduced a ban on water privatization in 2006; Bolivia, which by entering the right to water into the Constitution prohibited the privatization of water resources; and Slovenia, which changed its Constitution in order to respect the right to water and protect water resources (Eman et al., 2016, 2017; Humar & Eman, 2017).

A scarcity of clean drinking water is one of the number one issues of the twenty-first century because humans are dependent on water. Yet, "The water problem is not simply one of neglect and mismanagement. It is a problem actively constructed through harmful human intervention." (Whelan & White, 2005, p. 142). The lack of water is most often linked to the overexploitation of natural resources and as being the result of global warming (White & Heckenberg, 2014). Population growth could trigger a substantial global crisis related to water. Therefore, we must be even more aware of water crimes, and social and environmental harms associated with particular kinds of water ownership and management, because such acts have a considerable effect on the quality and quantity of water.

In many countries the supply of drinking water is enabled through water pipes and sewage systems. But it is sold and delivered in other ways as well. It has evolved to become a prized commodity, something valued for its exchange-value as well as its use-value. For instance, Brisman and South (2014) point to examples where entrepreneurs have deliberately not invested in local water systems for the delivery and treatment of drinking water and which would allow people to water from the taps, since they can make more profit from the sale of bottled water. Many companies started to sell bottled water because it is so financial lucrative, which is in a way very ironic. Bottled water is produced from spring water, to which is then added various ‘additives’ to ensure a longer shelf life. This means that the quality is worse than the original spring water, but the process also simultaneously increases the demand, since to produce 1 L of bottled water requires 3 L of spring water (Brisman & South, 2014).

The privatization of water differs from the privatization of most other resources because water is a good underpinning all life, and this is just one of the most worrying factors associated with the privatization of water (Todurov, 2014). In a world where water supplies are constantly decreasing, the transfer of the right to a water source to a private company can be very problematic, especially since there is a possibility that the company will work only for their own benefit and neglect the needs of all residents. Wisegeek (2014) presents the possibility of violations of human rights when private companies focus solely on profit. Examples drawn from France (Marn, 2013), India (Jayaraman, 2002), Bolivia (Finnegan, 2002), Greece (Petitjean, 2014), and South Africa (Bond, 2004) show how water privatization presents a particular form of crime against water insofar as the new owners engage in excessive depletion of water resources. This in turn contributes to changes in the quantity and quality of water and makes it immeasurably more expensive and therefore financially exclusive to only those who can afford it.

McClanahan (2014) observes that each year 5 million people die, of which the majority are children, due to the poor quality of drinking water or reduced access to clean drinking water. Part of this trend is due to privatisation, as discussed above. Putting water ownership and control in the hands of the private sector can seriously undermine human rights, including the human right to clean drinking water. As already mentioned, water is a prerequisite for the lives of people, non-human animals and plants as well as an indispensable resource for the economy; thereby the safeguarding of the water is at the utmost importance. This requires attention to matters such as controlling pollution, implementing water quality safeguard measures, maintaining watercourses and catchment areas, and implementing good agricultural practices.

The consequences of crime against water (including its privatization) can be catastrophic for all living creatures in the exposed area. Devastating effects are not only due to over-use. They also stem from pollution and contamination. An oil spill at one of Shell’s operations in Nigeria, for example, heavily polluted surrounding waters, which will remain polluted for decades to come (Jagger, 2014; UNEP, n.d.). A similar example is the contamination of water resources in Ecuador as the by-product of oil extraction undertaken by Chevron. Residents of the area were then forced to use contaminated water in their everyday use. In Bhopal, India it has been over 30 years since the release of dangerous chemicals into the nearby area. Water there is still contaminated.

In addition to the pollution a continuing problem is the over-exploitation of water resources. In India, huge amounts of water are used by farmers for irrigation of agricultural land. If there is no change, however, water resources will be totally depleted by 2050 (Karnani, 2014). Meanwhile, because of extensive agricultural irrigation affecting its main river tributaries, the Aral Sea is drastically lower and only a pale shadow of what it was a few years ago (Spoor, 1998). Its future is likewise in jeopardy.

Many writers are of the opinion that water may be the future cause of wars (Dunn, 2013; Gleick, 1993; Gupta, Dellapenna, & van den Heuvel, 2016). Changes in the nature of water supply can quickly lead to conflict. The United Nations (2015) point out that the sources of drinking water in the world are dramatically declining. The reasons for this are varied and include climate change, population growth and thus increased consumption, environmental crime such as pollution, and persistent depletion due to irrigation use (Agency of the Republic Slovenian Environment, 2015). Water scarcity now affects 40 per cent of the people in the world, and this number is constantly rising (UN, 2015).

Corruption also plays a significant role in water supply, production and distribution. For instance, petty bribes to government officials and water suppliers are used to falsify meter readings, avoid disconnections, and conceal illegal connections. The rising demand for water only serves to reinforce these corrupt practices and enforce a sense of discretionary power among officials, consequently weakening a country's rule of law (Global Initiative, 2019).

> As long as climate change, conflict and poverty continue to aggravate the Earth's dwindling water supply, water theft will remain a reoccurring and intensifying global problem. And while water theft remains viewed as a petty offense within a national context rather than a transnational organized crime issue, it is unlikely to be effectively resolved (Global Initiative, 2019).

Controlling water resources is a means of making money. Besides the issue of buying land with water springs or underground water reserves, the process of privatisation of water management in the cities opens the door to further opportunities for corruption. As emphasized by Alston and Mason (2008) the re-conceptualisation of water as a tradable commodity opens a Pandora box around water access and the rights of the users.

4 Water Crimes and Green Criminology

In a report on farming in Northern Australia for the Australian Agricultural Resource Economics Society, Campbell (2016, p. 17) notes that 'freshwater ecosystems are the most threatened globally'. Restricted quantities of clean water make it a particularly valuable property for those who own and control it. This points to the importance of analysing the conditions and circumstances underlying the theft of water (Global Initiative, 2014; Njeru, 2012), water-related corruption and violence, and water market price fixing (Kenya Water for Health Organisation, 2009, p. 27).

Water-related crimes are an emerging global issue (Interpol, 2016; Water Crimes Project, 2016). Identified research challenges include the absence of an international classification of water-related crimes; no working definition of water-related crimes; the scarcity of criminological studies; no mapping of criminal behaviour, motivation and threats; and complexity in data collection (Water Crimes Project, 2016, p. 6). Mattioli and Segato (2016) point out that water-related crimes are often recorded under other offences (e.g. fraud, corruption, trafficking, and falsification of documents due to lack of a systematic analytical approach leading to agreed definitions and international classification of water related crimes). Nonetheless, they define *water crime* as "any punishable contravention or violation of the limits on human behaviour as imposed by national criminal legislation, against surface, and ground water, or against water services" (Mattioli & Segato, 2016, p. 14). From this perspective, water crime includes any intentional act that presents potential harm or damage to water.

Examples of different types of water crimes identified by the European Union include pilfering of water from pipelines, illegal waste management, water theft, river and marine pollution, manipulation of sampling methods to avoid treatment costs, fraud and illegal trafficking of water, terrorism and cyber threats to water management operations, illegal waste discharges from factories, and unauthorised consumption from the water network. Another form of water crime is smuggling. For example, rich Canadian water resources are smuggled into the USA, where it is sold for large amounts of money; Kelly and Oldring (2015) describe a case where 50 two-litre bottles were sold for several hundred dollars. An Australian study by Barclay and Bartel (2015) on the perspectives of farmers, provides insight into how fresh water is being 'stolen' through diversion, despoiling and depletion, and how water theft results in availability and access injustices (Barclay & Bartel, 2015; Greiner, Fernandes, McCartney, & Durante, 2016). The unauthorised taking of surface or ground water has the cumulative impact of jeopardizing the sustainability of water resources. While awareness of environmental crimes is gaining momentum internationally, crimes against water are still an emerging issue. Such crimes have significant impacts, although intelligence is scarce, and information fragmented.

In light of what we presently know about water crimes and the potential drivers of increases in such crimes, there is a critical need for research in this area. Johnson et al. (2016, p. 160) suggest that green criminology and environmental protection should focus on "legal and governance frameworks that prioritise the human right to water and ecological sustainability over private interests". Besides this, a guarantee is needed for those countries in possession of abundant water resources that they will not become the targets of other countries or water conglomerates while, on the other hand, there is the need for a well-thought-out arrangement whereby these countries can help people living in water-scarce areas. Gleick (2006) recommends protecting water systems through a combination of improved physical barriers, more extensive real-time biological and chemical monitoring and treatment, and by developing smart and rapid integrated response strategies.

A big stumbling block to the equitable sharing of water and the protection of water supply is its commodification. Johnson et al. (2016, p. 149) stress that due to its scarcity and its importance for the basic survival, fresh water has become a highly prized asset and increasingly lucrative investment. Because water scarcity is presenting a huge profit opportunity the water industry is growing fast. Water conglomerates (e.g., Aquadust Alliance) are "creating databases that chart water supplies and identify risks and opportunities for business" (Johnson et al., 2016, p. 149). This is rather alarming, especially for the areas and countries that have rich water resources such as for example Slovenia and Tasmania. Water is thus treated solely as a commodity and an exploitable natural resource, not as a human right.

In expanding the field of water protection and water crime prevention, green criminology must in detail study and analyse the nature of water harms and water management (including enforcement, prosecution and sentencing practices) and the systems of criminal, civil and administrative law designed to manage, protect and preserve water. As this chapter has demonstrated, trends such as the privatization of water ownership and control have major negative implications for water human rights. Simultaneously, preventable environmental harms, such as climate change, are contributing to water scarcity and the pollution of existing water resources. A variety of crimes against water are emerging, many of which are predicted to increase in the near future. These are core issues of social and ecological justice. As such, they warrant greater attention from green criminology and indeed criminology more generally.

References

Alston, M., & Mason, R. (2008). Who determines access to Australia's water? Social flow, gender, citizenship and stakeholder priorities in the Australian water crisis. *Rural Society, 18*(3), 214–219.

Archer, J. (1998). *Sydney on Tap*. Sydney: Pure Water Press.

Barclay, E., & Bartel, R. (2015). Defining environmental crime: The perspective of farmers. *Journal of Rural Studies, 39*, 188–198.

Bennett, R. R., & Lynch, J. P. (1990). Does a difference make a difference?: Comparing cross-national crime indicators. *Criminology, 28*(1), 153–182.

Bond, P. (2004). *The battle over water in South Africa.* Accessed 18 May, 2017 from http://www.africafiles.org/article.asp?ID=4564

Bricknell, S. (2010). *Environmental crime in Australia, Research and Public Policy Series No. 109: Water theft* (pp. 103–114). Canberra: Australian Institute of Criminology.

Brisman, A., & South, N. (2014). *Green cultural criminology: Constructions of environmental harm, consumerism, and resistance to ecocide.* London: Routledge.

Brisman, A., McClanahan, W., & South, N. (2016). Water security, crime and conflict. In *Oxford Handbooks in Criminology and Criminal Justice*. Oxford: Oxford University Press. https://doi.org/10.1093/oxfordhb/9780199935383.013.86

Brisman, A., McClanahan, B., South, N., & Walters, R. (2018). *Water, Crime and Security in the Twenty-First Century: Too Dirty, Too Little, Too Much. Critical Criminological Perspectives.* London: Palgrave Macmillan.

Campbell, A. (2016). *Farming in northern Australia, where annual droughts, floods, fires & storms are guaranteed.* In Contributed presentation at the 60th AARES annual conference, Canberra, ACT, 2–5 February, 2016. Retrieved 30 May, 2016, from http://ageconsearch.umn.edu/bitstream/235260/2/campbell%20upload.pdf

Crank, J. P., & Jacoby, L. S. (2015). *Crime, violence, and global warming*. Abingdon: Routledge.

Dunn, G. (2013). Water wars a surprisingly rare source of conflict. *Harvard International Review, 19*(2), 46–49.

Eman, K., Kuhar, S., & Meško, G. (2016). Providing water supply in local communities—Case study of Slovenia. In G. Meško & B. Lobnikar (Eds.), *Criminal justice and security in central and eastern Europe. Safety, security, and social control on local communities* (pp. 367–374). Faculty of Criminal Justice and Security: Ljubljana.

Eman, K., Kuhar, S., & Meško, G. (2017). Water crimes and policing. *Varstvoslovje, 19*(2), 105–119.

Eman, K., Kuhar, S., & Meško, G. (2018). Kriminaliteta zoper vodo v Sloveniji. *Revija za kriminalistiko in kriminologijo, 69*(1), 44–53.

Eman, K., & Meško, G. (2013). Green issues in South-Eastern Europe. In N. South & A. Brisman (Eds.), *Routledge international handbook of green criminology* (pp. 230–240). London/New York: Routledge.

Eman, K. & Meško, G. (2020). The access to safe and affordable drinking water as a fundamental human right—The case of the Republic of Slovenia. In J. Blaustein, K. Fitz-Gibbon, N. Pinto & R. White (Eds.), *Emerald Handbook of Crime, Justice and Sustainable Development*. London: Emerald.

Finnegan, W. (8. 4. 2002). Leasing the rain. *The New Yorker*. Accessed 18 May, 2017 from http://www.newyorker.com/maga-zine/2002/04/08/leasing-the-rain

Gleick, P. H. (1993). *Water in crisis: A guide to the world's fresh water resources*. Oxford: Oxford University Press.

Gleick, P. H. (1998). Water in crisis: Paths to sustainable water use. *Ecological Applications, 8*(3), 571–579.

Global Initiative (2014). *Water, water everywhere: charting the growth of organized water theft, in analysing organised crime.* Accessed 18 May, 2017 from https://www.globalinitiative.net

Global Initiative (2019). *Water, water everywhere? Charting the growth of organized water theft.* Retrieved 2 April, 2017, from http://globalinitiative.net/water-smuggling/

González Rivas, M. (2014). Ethnolinguistic Divisions and Access to Clean Water in Mexico. *Latin American Research Review, 49*(2), 129–151.

Greiner, R., Fernandes, L., McCartney, F., & Durante, J. (2016). Reasons why some irrigation water users fail to comply with water use regulations: A case study from Queensland. *Land Use Policy, 51*, 26–40.

Gupta, J., Dellapenna, J. W., & van den Heuvel, M. (2016). Water sovereignty and security, high politics and hard power: The dangers of borrowing discourses! In C. Pahl-Wostl, A. Bhaduri, & J. Gupta (Eds.), *Handbook on water security* (pp. 120–136). Cheltenham: Edward Elgar.

Humar, S., & Eman, K. (2017). Privatizacija vodnih virov – nova oblika kriminalitete zoper vode. *Revija za kriminalistiko in kriminologijo, 68*(2), 116–134.

Interpol (2016). *Environmental crime*. Retrieved 2 April, 2017, from www.interpol.int/Crime-areas/Environmental-crime/Environmental-crime

Jayaraman, N. (2002). *No water? Drink Coke!* India Resource Center. Accessed 18 May, 2017 from http://www.indiaresource.org/campaigns/coke/2003/nowaterdrinkcoke.html

Jagger, B. (2014). Crimes Against Present and Future Generations. *Challenge, 57*(1), 41–59.

Johnson, H., South, N., & Walters, R. (2016). The commodification and exploitation of fresh water: Property, human rights and green criminology. *International Journal of Law, Crime and Justice, 44*, 146–162.

Karnani, A. (2014). Corporate social responsibility does not avert the tragedy of the commons. Case study: Coca-Cola India. *Economics, Management, and Financial Markets, 9*(3), 11–33.

Kelly, P., & Oldring, P. (2015). *Meet a Canadian making thousands of dollars smuggling water into the US.* Retrieved 2 April, 2017, from https://www.cbc.ca/radio/thisisthat/school-real-estate-word-search-champion-water-smugglers-1.3194224/meet-a-canadian-making-thousands-of-dollars-smuggling-water-into-the-u-s-1.3194276

Marn, U. (1. 2. 2013). Plenilska doktrina. *Mladina*. Accessed 18 May, 2017 from http://www.mladina.si/120104/plenilska-doktrina/

Mattioli, W., & Segato, L. (2016). *Environmental crimes in the water sector. Water crimes project.* Accessed 18 May, 2017 from http://www.watercrimes.eu/#resources

McClanahan, B. (2014). Green and Grey: Water Justice, Criminalization, and Resistance. *Critical Criminology, 22*(3), 403–418.

Meško, G., & Eman, K. (2012). Organised crime involvement in waste trafficking - case of the Republic of Slovenia. *Kriminalističke teme: časopis za kriminalističku teoriju i praksu, 12*(5/6), 79–96.

National Geographic. (2016). *Freshwater crisis.* Retrieved 27 March, 2017, from http://environment.nationalgeographic.com/environment/freshwater/freshwater-crisis/

Njeru, G. (2012). *Water shortages driving growing thefts, conflicts in Kenya.* London: Thomson Reuters Foundation.

Pečar, J. (1981). Ekološka kriminaliteta in kriminologija. *Revija za kriminalistiko in kriminologijo, 34*(1), 33–45.

Petitjean, O. (28. 3. 2014). *Forced privatizations in Greece: Suez eyes taking over the water services of Athens and Thessaloniki.* Multinationals Observatory. Accessed 18 May, 2017 from http://multination-ales.org/Forced-Privatizations-in-Greece

South, N., & Brisman, A. (2013). *Routledge international handbook of green criminology.* London, New York: Routledge.

Spoor, M (1998). The Aral Sea Basin Crisis: Transition and Environment in Former Soviet Central Asia. *Development and Change, 29*(3), 409–435.

Todurov, M. (24. 9. 2014). *Privatizacija 2.0. Vzpon.* Accessed May 18, 2017 from http://vzpon.vzajemci.com/novice/gospodarstvo/11603/priva-tizacija_2_0

United Nations [UN]. (2015). *The millennium development goals report 2015.* Accessed 18 May, 2017 from http://www.un.org/mil-lenniumgoals/2015_MDG_Report/pdf/MDG%202015%20Summary%20web_english.pdf

United Nations Department for Economic and Social Affairs [UNDESA]. (2014). The human right to water and sanitation. Accessed 18 May, 2017 from http://www.un.org/waterforlifedecade/human_right_to_water.shtml

Water Crimes Project (2016). *Water crimes.* Retrieved 27 March, 2017, from http://www.watercrimes.eu/

Whelan, J., & White, R. (2005). Does privatising water make us sick? *Health Sociology Review, 14*(2), 135–145.

Whelan, J. J., & Willis, K. (2006). Problems with provision: barriers to drinking water quality. *Journal of Water Supply: Research and Technology, 55*(1), 3–10.

White, R. (1998). Environmental criminology and Sydney water. *Current Issues in Criminal Justice: Journal of the Institute of Criminology, 10*(2), 214–219.

White, R. (2003). Environmental issues and the criminological imagination. *Theoretical Criminology, 7*(4), 483–506.

White, R. (2016). Building NESTs to combat environmental crime networks. *Trends Organised Crime, 19*(1), 88–105.

White, R., & Heckenberg, D. (2014). *Green criminology: An introduction to the study of environmental harm.* London: Routledge.

White, R., & Perrone, S. (2015). *Crime, criminality & criminal justice* (2nd ed.). Victoria: Oxford University Press.

Wisegeek. (2014). *What is water privatization?* Accessed 18 May, 2017 from http://www.wisegeek.com/what-is-water-privatization.html

World Health Organization [WHO]. (2015). Progress on sanitation and drinking water: 2015 Update and MDG Assessment. Accessed 18 May, 2017, from http://www.wssinfo.org/fileadmin/user_upload/resources/JMP-Update-report.html

Chapter 2
The Politics of Water Rights: Scarcity, Sovereignty and Security

Avi Brisman, Bill McClanahan, Nigel South, and Reece Walters

1 Introduction

The city of Cape Town, South Africa, owes its origins to a settlement established by the Dutch East India Company in 1652. The fresh water running down from Table Mountain meant irrigation and that, as a result, vegetables could be grown and sheep and cattle could graze. Today, in 2017 and 2018, Cape Town's water supply is no longer plentiful with reduced rainfall unable to satisfy increased demand as urban growth continues. Supply has been dangerously low (Kane, 2018; McGrath, 2018; McKenzie & Swails, 2018; Pérez-Peña, 2018; Welch, 2018). The city experienced its driest year to that point in 2017, but if water levels continue to fall in 2018, city officials will take the unprecedented step of declaring 'Day Zero' when '[t]aps in homes and businesses will be turned off until the rains come. The city's four million residents will have to line up for water rations at 200 collection points' (Onishi & Sengupta, 2018, p. A1).

A. Brisman
School of Justice Studies, Eastern Kentucky University, Richmond, KY, USA

School of Justice, Queensland University of Technology, Brisbane, QLD, Australia

Newcastle Law School, University of Newcastle, Newcastle, NSW, Australia
e-mail: Avi.Brisman@eku.edu

B. McClanahan (✉)
School of Justice Studies, Eastern Kentucky University, Richmond, KY, USA
e-mail: Bill.Mcclanahan@eku.edu

N. South
Department of Sociology, University of Essex, Colchester, UK
e-mail: n.south@essex.ac.uk

R. Walters
School of Humanities and Social Sciences, Deakin University, Burwood, VIC, Australia
e-mail: reece.walters@deakin.edu.au

K. Eman et al. (eds.), *Water, Governance, and Crime Issues*,
https://doi.org/10.1007/978-3-030-44798-4_2

At the time of this writing (April 2018), the dreaded 'Day Zero' had been postponed several times largely because of restrictions on personal and agricultural usage, reflecting the resiliency of Cape Town's residents (McKenzie & Swails, 2018). But the threat looms—and serves as a warning to other cities, especially as climate change portends an increase in extreme weather events, such as more frequent droughts (McKenzie & Swails, 2018; Welch, 2018). Indeed, the same combination of overdevelopment, population growth and climate change—exacerbated by inadequacies in infrastructure—has upset the balance between water use and supply, augmenting the risk of severe drinking-water shortages in urban centres from North America to South America, Asia to Australia (Welch, 2018).

While the situation in Cape Town represents a 'doomsday scenario' for many of the city's residents, McKenzie and Swails (2018) point out that for the millions of 'Capetonians' who live in 'vast shack lands [and] use less than 5% of the city's water'—in comparison to the more than 60% consumed by people in houses and apartments—'every day is already like Day Zero'. Unfortunately, such massive inequalities in South African and elsewhere—a reflection of colonial legacies—are likely to become more pronounced with growing water scarcity. Indeed, as Jackson and colleagues (Jackson & Sultana, 2016, p. 2) write in their introduction to *Eating, Drinking: Surviving*:

> the combined effects of climate change (increasing extreme weather events such as floods and droughts), the pressures of population growth and urbanization on the availability of agricultural land, and the increasing global demand for meat and animal products, ... are threatening to undermine the ability of current agri-food systems to feed a global population which it is estimated will increase from around 7.4 billion today to around 9.6 billion by 2050 (UN, 2013). The need for sufficient clean water to meet the needs of both thirsty people and crops have also put enormous pressure on dwindling water resources, creating water crises in drinking water systems and irrigation systems, bringing into question how we manage surface and ground water sources. To cope with future challenges innovative solutions have to be sought. Within the same time period it is also estimated that there will be a quadrupling in the global economy, a doubling in the demand for food and fuel, and a more than 50% increase in the demand for clean water (Foresight, 2011).

It should be clear, then, that because water is such an essential component for human and nonhuman life and ecological stability, lack of access to clean fresh water (partly caused by harms, such as pollution) represents a significant risk to humanity's future. It follows then that harms to water, whether anticipated or realized, often contribute to conflict and create social instability (and vice versa). In this chapter, we describe some of the ways in which threats to a balanced global hydroecology represent challenges for the health and vitality of human populations and human security.[1] Our focus is on five major (overlapping) issues of concern: availability, pollution, privatisation, security and conflict. As we attempt to demonstrate, these areas are interconnected—much like streams and rivers themselves—and the extent

[1] We employ the language of 'security' here while also recognizing that, in many ways, it is a discourse and technology of the very capitalist logics and architectures that so often create harms to hydroecologies in the first place (McClanahan & Brisman, 2015; Neocleous, 2008; Neocleous & Rigakos, 2011; Wall, 2013).

to which these linkages are acknowledged (as well as when and where) may well influence equitable access to clean fresh water. We begin by describing some of the problems of water pollution and the (often associated) problems of violations of water sovereignty that characterize many historical and contemporary water conflicts using the examples of Standing Rock, in North Dakota throughout 2016, and the ongoing (as of this writing) conflict over Oregon's Klamath River. Next, we turn our attention to issues of water privatisation—and the conflicts implicated in privatisation schemes—in Bolivia, Ireland, and Michigan. Finally, we address some increasingly common concerns over climate and drought-induced migration, giving special attention to the ways in which the spectre of the 'climate refugee' configures contemporary debates over water through the capitalist logics of security and securitization.

2 Water Pollution and Sovereignty: Conflict and Resistance at Standing Rock and Elsewhere

Fresh water, in addition to being essential for the healthy and balanced functioning of biotic life, represents a significant opportunity for exploitation at the hands of global capitalist modes of economic organization (Brisman & South, 2017; Brisman, McClanahan, South, & Walters, 2018). Because of the necessity of water, it is perhaps the ultimate commodity for the capitalist imagination: along with food and air, water is vital for life, and so unlike other marketable goods (e.g., consumer electronics) and services (e.g., entertainment products and media), it is impossible to live without access to clean drinking water. It is little surprise, then, that water takes a central place in the markets of global capitalism. Indeed, in a 2018 speech to attendees of the annual Water UK City Conference, UK Environmental Secretary Michael Gove characterized the water business as one that enjoys 'a captive set of consumers, guaranteed income from hundreds of thousands, if not millions of customer accounts, and a certainty that your product will never go out of fashion' (Gove, 2018). The Environment Secretary's comments endorse what hedge fund managers and investment bankers have been saying for many years, that the corporatization of fresh water is the next 'big super trend' for commercial profits (Prosser, 2016) What results from this marketization and commodification of water are, then, often the marginalization and exploitation of human populations and communities. The privatisation of fresh water has resulted in the widespread exploitation of the world's poorest and most vulnerable people in the global south. The need to reduce the global 'water footprint' from the corporate trade monopolies of multinational enterprises is essential for recognizing the human right to water (Bhattacharya & Banerjee, 2015). The harms associated with water in these examples can come from conflicts over rights to water that are sometimes conditioned by claims to water sovereignty relating to cultural and spiritual practice. The first example we offer of such harms is that of the conflicts over the Dakota Access Pipeline (DAPL).

The DAPL is a nearly 1200 mi-long pipeline designed to carry oil from the Bakken oil fields in North Dakota to holding tanks in Illinois. Because the pipeline crosses significant swaths of land sacred to Native American tribes (including the Meskwaski and Sioux nations) and because of tribal concerns over the disruption and possible pollution of rivers and tribal burial grounds, the proposed pipeline was met with fierce opposition prior-to and during its construction in 2016. After months of sustained protests that included direct action to block construction, solidarity marches across the country and world, and sometimes violent clashes with police, pipeline security, and construction workers, the permits necessary to continue were temporarily denied by the Army Corp of Engineers (Samson & Ravna, 2016). Even as it became increasingly clear that the protestors (called 'water protectors') would enjoy a temporary victory in blocking the pipeline, a similar pipeline carrying oil across rural North Dakota, the Belle Fourche pipeline, leaked nearly 200,000 gallons of crude oil into the Ash Coulee creek and an adjacent hillside. The spill highlights the concerns of the DAPL protestors, of course, but it also boldly underlines what are the essential concerns of this section: water is increasingly under threat, by the dominant global logics and relations of neoliberal capitalism and the repressive states that continue to support activities that pollute water, restrict access, and repress movements for water justice. Furthermore, those threats will often—and, perhaps, increasingly—lead to conflict. That the DAPL protests—perhaps the most attention-grabbing social resistance movement of 2016—were in direct response to anticipated and realized water injustice further highlights the increasing likelihood of social unrest and instability flowing from growing and sustained water injustice.

Throughout the months of conflict surrounding the construction of the pipeline, there were countless significant moments of repression and state violence carried out at the hands of authorities and private security actors at the behest of the interests of state-corporate petro capital. Protestors at Standing Rock, as the resistance camps came to be known, were routinely teargassed, 'kettled', beaten, subject to intense surveillance, and shot with so-called 'less-than-lethal' crowd control weapons by police. In addition to the violence of those official authorities, protests were surveilled and disrupted and individual protestors were attacked by members of TigerSwan, a private security firm contracted by Dakota Access, LLC, and the corporate group constructing the pipeline. In documents obtained by media outlets, it was revealed that TigerSwan engaged in 'military-style counterterrorism measures' in order to disrupt protests and sway public opinion in favour of pipeline construction (Brown, Parrish, & Speri, 2017; May, 2016; Samson & Ravna, 2016).

Despite the temporary victories of the movement to block construction of the DAPL in 2016, in January 2017, shortly after entering office, President Donald J. Trump reversed previous executive decisions by issuing a memorandum approving the final construction of the pipeline. With the last protestors evicted from the camps at the Standing Rock campground evicted (with ten arrested), the DAPL was completed in April of 2017 (Jones, Diamond, & Kreig, 2017; Renshaw, 2017). In the events surrounding the construction of the DAPL, we can locate not only the ways in which conflicts over water that relate to concerns over rights, sovereignty, and pollution can lead to state-sponsored and enacted violence, but also to the sorts of social instability and division that increasingly characterize a world connected by the global logics of capitalism.

At the time of this writing—April 2018—"the next Standing Rock" is unfolding in Oregon (Gentry & Marris, 2018). The Klamath River, which runs from Oregon through northern California, emptying in the Pacific Ocean, is vital for anadromous fish migration and sacred to a number of Native American tribes. With the planned decommission of four hydroelectric dams, the Klamath River is set to undergo one of the largest river restoration projects in American history, opening up more spawning and rearing habitat for salmon and improving overall water quality (Gentry & Marris, 2018; Stecker, 2016). The prospects of a revitalized river are being threatened, however, by the proposed Pacific Connector Gas Pipeline, which would run from Malin to Coos Bay, Oregon, crossing under the Klamath River (Gentry & Marris, 2018). If the pipeline gets built, it would extend the Ruby Pipeline, which runs from Opal, Wyoming, to Malin, Oregon, enabling natural gas in to a storage and export terminal for sale to countries in the Pacific Rim. As is often the case, the company behind the pipeline (Pacific Connector Gas Pipeline LP) and export terminal (Jordan Cove Energy Project LP)—the Pembina Pipeline Corporation—is promising that the pipeline and terminal will create jobs and generate tax revenue for some of the poorest areas in Oregon, while ensuring that the pipeline will be safe (Gentry & Marris, 2018). As Gentry and Marris (2018) report, however,

> The 95-foot-wide gash through the tribes' ancestral territory that pipeline construction would require would be likely to unearth long-buried ancestors and pulverize sites of cultural importance. Construction would strip shade from streams and pollute them with sediment, harming fish central to the Klamath's traditions and way of life. If the pipeline catches fire or leaks, the Klamath River and its fish will be put at risk. The track record of fossil fuel pipelines suggests such a calamity is only a matter of time.

To be fair, natural gas does generate lower greenhouse gas emissions than coal. But it is still a fossil fuel and leaks during extraction and transport, as well as the burning of natural gas, all contribute to climate change. Here, again, we see the linkages between water-related harms (or the potential thereof) and climate change. Given the threat that the pipeline poses to the river (its aquatic life, its watershed), as well as to the Klamath people's homes and cultural and spiritual ways of life, it is unsurprising that the original stewards of this area, along with concerned citizens and conservation groups, have dug in their heels for a fierce fight (Gentry & Marris, 2018; Sorisingtide, 2019).

3 Privatisation and Resistance: Bolivia and Ireland

In addition to the sorts of conflict surrounding concerns about violations of traditional water sovereignty and the pollution of culturally and spiritually significant water sources such as those described above, the privatisation of water and water provision also creates significant conflict. In this section, then, we offer some examples of cases in which water privatisation has contributed directly to water injustice, pollution, and conflict.

In 1998, the government of Bolivia, under the supervision of the World Bank, passed laws that effectively privatised the water supply system of Cochabamba, a region that over one million Bolivians call home, 60% of them indigenous. Concessions to manage—and profit from—the water system were granted to Bechtel, a multi-billion dollar global engineering and construction corporation. Bechtel, upon taking control of the Cochabamba water supply, immediately tripled prices and cut off water services to those unable to pay, going so far as to charge for rainwater gathered in homemade and traditional catchment systems. In response, *La Coordinadora de Defensa del Agua y de la Vida* (the Coalition in Defense of Life and Water) formed, quickly organizing a referendum that demanded the cancellation of all Bolivian water contracts with Bechtel. When the Bolivian authorities refused to back down, protestors took to the streets and were met with repressive violence from the state military, leaving dozens wounded and one seventeen-year-old protestor, Hugo Banzer, dead. The Bolivian government eventually capitulated, severing all contracts with Bechtel (Barlow, 2008, pp. 102–106. See generally: McClanahan, 2014). Bolivia's rejection of Bechtel's privatisation plans, and Morales' forceful and succinct reiteration of water's status as a common good that must remain in the commons, highlight the success of Bolivia's water movement in resisting the structural adjustments frequently demanded by global capital.

While privatisation schemes, frequently mandated as part of structural adjustment requirements attached to international financial aid and capital lending, are prevalent—particularly in the Global South—they do not represent the only impetus for the gestation of resistance. In the Global North, which has its own water issues, there are also meaningful pockets of intense and concerted resistance to the various forces that threaten access to useable water, as well as individual and community responses to drought and pollution events that jeopardize water quality and access. Following the global financial crises of 2008, with the Republic of Ireland's economy suffering huge losses and contraction, the IMF and EU offered a 'bailout' package in the form of loans carrying mandatory economic adjustments. Among those adjustments was the requirement that Bord Gáis, the nation's public water provider, sell off its customer supply business to private interests (Hearne, 2015). Accepting the terms of the bailout, Ireland sold off the provision of consumer water to a tripartite consortium consisting of Centrica and Brookfield Renewable Energy, multinational corporations heavily involved in global water and energy privatisation, and ICON Infrastructure, an investment capital group focused on finding profits in global infrastructure assets. The resulting company, Irish Water, immediately undertook plans to institute water charges through the strict metering of residential water use, leading to Ireland's own 'water rebellion' (O'Doherty, 2015). Just as in Bolivia, those most affected by the privatisation of water mobilized in protest, engaging in direct action activism and vocal dissent against Irish Water. Many of Irish Water's customers physically blocked the installation of new meters, while one 2014 demonstration against the new water charges saw participation from over 130,000 citizens, roughly 4% of the country's population, and national polls revealed that a full third of water uses planned to boycott the new charges (O'Toole, 2014).

Following the Irish general elections in 2016, during which intense campaigns focused heavily on water issues raised by Irish Water's privatisation efforts (Irish Press Association, 2016; Ryan, 2016), water charges of the sort rejected by protestors were suspended, with an expert commission recommending that normal and quotidian household water usage be paid by the state, with 'overuse' resulting in charges under a 'polluter pays' model (O'Halloran, 2016; RTE News, 2016).

4 Water and Security: Climate Change, Migration and Conflict

While the water conflicts described above illustrate the likelihood of social instability in the wake of efforts to privatise water provision, or to undertake pipeline construction in the name of global extractive capitalism, there are also new sources of water conflict emerging on the horizon. With the ongoing advance of climate change, global hydro-ecology is changing at a rapid rate, with droughts, floods, and rising sea levels each potentially leading to increased social instability, water insecurity, and conflict. While the examples above illustrate 'conflicts' between state-corporate interests and popular movements resisting the realized or potential harms to water represented by unjust pollution and privatisation of water, the examples offered in this section illustrate the various ways in which the hydro-ecological effects of climate change will potentially lead to social conflict, insecurity, and instability.

As Welch (2018) argues, 'droughts in recent years have helped spark famine and unrest in rural nations around the Arabian Sea, from Iran to Somalia'. Perhaps the most salient contemporary example of the ways in which climate change's impact on water can be a potential source of social conflict and instability is that offered by the ongoing war in Syria. Starting in 2011, the civil war in Syria has been a conflict between state forces and armed militias loyal to President Bashar al Assad and armed forces representing a range of groups including Islamic State (ISIS/ISIL) forces, Kurdish independence fighters, Sunni rebel groups such as the Free Syrian Army, and more. While the origins of the conflict are too complex to cover here, there appears to be a broad consensus among analysts that climate change (and, by way of climate change, water) has played a significant role (Gleick, 2014; Selby, Dahi, Fröhlich, & Hulme, 2017a, 2017b). The conflict, indeed, has frequently been referred to as a 'climate change war' (sometimes garnering mentions alongside the conflict in Darfur, Sudan as 'the *first* climate change war') (Welzer, 2012). At the core of the relationship between water, climate change, and the conflict in Syria is the migration of thousands of rural people into the urban core of Syrian cities like Damascus, with the argument being, essentially, that climate change-induced droughts in the rural 'breadbasket' of Syrian farmlands pushed populations into the cities, where social tensions and competition for space and resources further fuelled the basis for conflict.

The explanation that climate change and drought were pivotal in tilting Syria into civil war finds support in green criminological analyses of the possible criminogenic effects of climate change. As Agnew (2012) has described, among the anticipated criminogenic forces considered in this stream of criminological analysis and 'horizon scanning' are 'a number of crimes at the individual, corporate, and state levels' (Agnew, 2012, p. 26) including individual acts of theft and violence, corporate crimes of bribery and pollution, and increased levels of criminal state-level aggression. Applying the theoretical frameworks provided by general strain theory, Agnew develops a model of climate change and crime, noting that some effects of climate change—habitat change, food and water shortages, forced migration, and loss of livelihood—often precede and contribute to increased levels of strain experienced at every level of society. These strains are likely to lead to the strengthening of 'criminogenic mechanisms' including reduced social support and increased social conflict. Fussey and South (2012) introduce problems of environmental (in) security arising from climate change (for a further discussion of the links between climate change, security/insecurity, and securitization, see McClanahan & Brisman, 2015; Brisman et al., 2018), describing the ways that climate change is likely to cause conflicts beyond the scale of the individual and community level conflicts addressed by Agnew, extending instead to the state level. As Fussey and South (2012) note, the advance of anthropogenic climate change will cause resource shortages, with those shortages likely leading to conflicts between nation states over water, land, and other essential components of human and nonhuman life.

What a green criminology focused on the exploitation of water and people must be concerned with, though, are the potential problems of assuming that migration will necessarily lead to conflict. Narratives that suggest that conflicts such as the ongoing war in Syria are linked to climate migration, and that migration will lead to conflict, risk advancing populist and racist or xenophobic notions of the dangerous 'other'. As Eklund and Thompson (2017) note, historical links between conflict and migration are rare, with migration sometimes strengthening or improving social and economic conditions in the receiving country or region. This is not, of course, to say that climate change-induced drought and flooding will not continue to affect migration; it is already a realized impact of climate change—one that is likely to increase. There is little evidence, however, to suggest that migration will *inherently* or *necessarily* lead to conflict, and to suggest that it will risks 'overlooking important socio-political factors' (Eklund & Thompson, 2017; Tertrais, 2011). In the panic swirling around environmental 'push factors' like drought-induced migration, we can locate—as is often the case—the spectre of (environmental) 'security'. 'Environmental security' has been defined in various ways (for discussions, see, e.g., Brisman, McClanahan, & South, 2016; South, 2012), but for our purposes, we note Hecker's (Hecker, 2011, p. 12) notion of 'environmental security' as '[t]he current and future availability (determined by the factors—supply, accessibility and management) of life-supporting ecosystem services and goods for human needs and natural processes which contribute to poverty alleviation and conflict deterrence'.

While much of the drive for water security reflects concern for quotidian human uses (e.g., consumptive, agricultural, industrial), there are also water security initiatives that reflect the militaristic discursive undertones of 'security' (Molotch, 2012; Neocleous & Rigakos, 2011; Neocleous, Rigakos, & Wall, 2013; Stavriankas & Selby, 2013) endemic to the production and reproduction of racist, nativist, and nationalist capitalist order. It seems, then, as though many calls for 'water security' are part of the various and routine ways in which the languages and technologies of 'security' are deployed in efforts to strengthen states, regimes, and borders, efforts which are often an expression of the racism and xenophobia that drive and complement global capitalism (Brisman et al., 2018; Brisman & South, 2017). Related to this point, and to our broader discussion of the linkages between water, migration, and conflict, Meinen (2017) suggests that the figure of the 'climate refugee' exists purely in the social imagination, finding its only 'empirical quality…in the various future-conditional knowledge practices that produce it' (Gemenne & Baldwin, 2013, p. 267). In many ways, then, the popular image of 'the climate refugee' is as described by Meinen: a 'speculative' fiction deployed in order to further the aims of 'security' required for ongoing global capitalism, with the potential of creating a 'neo-police consisting of NGOs repurposed by a state that is labouring to translate' the 'humanitarian "concern"' of climate-induced migration into a 'security "problem"' (Meinen, 2017).[2]

5 Conclusion

There are significant reasons to locate the origins of certain conflicts in the problems of hydro-ecology, as described in the previous examples of Standing Rock, Ireland and Bolivia, and Syria. We must, however, be mindful of the dangers in reducing our analysis of those conflicts in ways which prop up or advance growing narratives of racist and xenophobic populist politics. As Klein (2016, p. 13) notes, contemporary trends toward the violent 'othering' of the dispossessed and displaced are worrying. Given the lack of progress in preparing for the consequences of a warming world, she warns that prevailing logics and the dominant culture of the white Global North seem likely to be 'willing to let the countries where black and brown people live disappear beneath the waves or desiccate in the arid heat'.

While the final outcome of the conflicts in Syria, the hydro-future of native lands in areas like North Dakota and Oregon, and the conditions faced by water-consumers in privatised systems like those in Ireland and Bolivia remains to be seen, what we do know now is that water will likely remain a significant site and source of local,

[2] Meinen's description of the securitization of climate change (McClanahan & Brisman, 2015) as potentially creating a 'neo-police' highlights an important point: police is the essential technology of a 'security' that functions to preserve capitalist social order through the accumulation of capital and the pacification of human populations and ecologies (Correia & Wall, 2018; Neocleous, 2008, 2014).

regional, and global conflict. The nature of and extent to which current and future 'water wars' occur remains to be seen, however. Conservation and conscious (and *conscientious*) consumption—not just of water, itself, but of food and clothing and gas and technology that require water for their production—will need to play a part (Zachos, 2017), but action is also needed to address the ways in which social and economic systems deny the 'status of water as a public good' and, in pursuit of profit, allow the 'hyper-commodification of the ultimate inelastic good' (Graham, Desai, & McFarlane, 2016, p. 134). The conditions of hydrological apartheid existing between the water-rich and water-poor in various geo-social contexts, lead Graham et al. (2016, p. 134) to argue for urgent challenge to the current state of the politics of water rights. Without equitable adjustment to the future distribution of water access, the consequences will be more 'water wars' and accompanying dehydration, disease, desperation and ultimately death.

References

Agnew, R. (2012). Dire forecast: A theoretical model of the impact of climate change on crime. *Theoretical Criminology, 16*(1), 21–42.

Barlow, M. (2008). *Blue covenant*. New York: The New Press.

Bhattacharya, S., & Banerjee, A. (2015). Water privatization in developing countries: Principles, implementations and socio-economic consequences. *World Scientific News*, Retrieved May 13, 2019, from http://www.worldscientificnews.com/wp-content/uploads/2015/06/WSN-10-2015-17-31.pdf

Brisman, A., McClanahan, B., & South, N. (2016). Water security, crime and conflict. In *Oxford handbooks in criminology and criminal justice*. Oxford: Oxford University Press. https://doi.org/10.1093/oxfordhb/9780199935383.013.86

Brisman, A., McClanahan, B., South, N., & Walters, R. (2018). *Water, crime and security in the twenty-first century: Too dirty, too little, too much*. London: Palgrave Macmillan.

Brisman, A., & South, N. (2017). Food, crime, justice and security: (Food) Security for whom? In I. Werkheiser & Z. Piso (Eds.), *Food justice in US and global contexts: Bringing theory and practice together* (pp. 185–200). New York: Springer.

Brown, A., Parrish, W., & Speri, A. (2017). Leaked documents reveal security firms counterterrorism tactics at standing rock to defeat pipeline insurgencies. *The Intercept*, 27 May 2017. Retrieved 14 May, 2019, from https://theintercept.com/2017/05/27/leaked-documents-reveal-security-firms-counterterrorism-tactics-at-standing-rock-to-defeat-pipeline-insurgencies/

Correia, D., & Wall, T. (2018). *Police: A field guide*. New York: Verso Books.

Eklund, L., & Thompson, D. (2017). Is Syria really a climate war? *The Conversation*, 21 July 2017. Retrieved 14 May, 2019, from https://theconversation.com/is-syria-really-a-climate-war-we-examined-the-links-between-drought-migration-and-conflict-80110

Foresight. (2011). *The future of food and farming: Final report*. London: Government Office for Science.

Fussey, P., & South, N. (2012). Heading toward a new criminogenic climate: Climate change, political economy and environmental security. In R. White (Ed.), *Climate change from a criminological perspective* (pp. 27–40). New York: Springer.

Gemenne, F., & Baldwin, A. (2013). The paradoxes of climate change and migration. In ISSC (Ed.), *World social science report: Changing global environments*. Paris: UNESCO.

Gentry, D., & Marris, E. (2018). The next standing rock? A pipeline battle looms in Oregon. *The New York Times*. March 8. Retrieved 13 May, 2019, from https://www.nytimes.com/2018/03/08/opinion/standing-rock-pipeline-oregon.html

Gleick, P. H. (2014). Water, drought, climate change, and conflict in Syria. *Weather, Climate, and Society, 6*(3), 331–340.

Gove, M. (2018). A water industry that works for everyone. *Speech Delivered to the 2018 Water UK City Conference*, 1 March 2018. Retrieved 14 May, 2019, from https://www.gov.uk/government/speeches/a-water-industry-that-works-for-everyone

Graham, S., Desai, R., & McFarlane, C. (2016). De-networking the poor: Revanchist urbanism and hydrological apartheid in Mumbai. In O. Coutard & J. Rutherford (Eds.), *Beyond the Networked City* (pp. 114–137). Abingdon: Routledge.

Hearne, R. (2015). Why privatizing state assets is not in the public interest. *Irish Examiner*, 26 January 2015. Retrieved 13 May, 2019, from https://www.irishexaminer.com/viewpoints/analysis/why-privatising-state-assets-is-not-in-the-public-interest-308850.html

Hecker, J. H. (2011). *Peace and sustainable development through environmental security: A methodology for environmental security assessments*. The Hague: Institute for Environmental Security.

Irish Press Association. (2016). Election candidates join thousands in Dublin protest against water charges. *The Irish Independent*, 20 February 2016. Retrieved 14 May, 2019, from https://www.independent.ie/breaking-news/irish-news/election-candidates-join-thousands-in-dublin-protest-against-water-charges-34470717.html

Jackson, P., Spiess, W. L. E., & Sultana, F. (2016). Introduction: Understanding the complexities of eating, drinking, and surviving. In P. Jackson, W. E. L. Spiess, & F. Sultana (Eds.), *Eating, drinking: Surviving. The International Year of Global Understanding—IYGU* (pp. 1–12). Cham: Springer.

Jones, A., Diamond, J., & Kreig, G. (2017). Trump advances controversial oil pipelines with executive action. *CNN News*, 24 January 2017. Retrieved 13 May, 2019, from https://www.cnn.com/2017/01/24/politics/trump-keystone-xl-dakota-access-pipelines-executive-actions/index.html

Kane, D. (2018). Running dry in Cape Town. *The New York Times*. February 2: A23. Retrieved 13 May, 2019, from https://www.nytimes.com/2018/02/01/opinion/cape-town-drought-day-zero.html

Klein, N. (2016). Let them drown: The violence of othering in a warming world. *London Review of Books, 38*(1), 11–14.

May, C. (2016). A reporter's notebook: The journey to Oceti Sakowin, the protest camp of the Standing Rock Sioux. *Salon* 25 December 2017. Retrieved 13 May, 2019, from https://www.salon.com/2016/12/25/a-reporters-notebook-the-journey-to-oceti-sakowin-the-protest-camp-of-the-standing-rock-sioux/

McClanahan, B. (2014). Green and grey: Water justice, criminalization, and resistance. *Critical Criminology, 22*(4), 403–418.

McClanahan, B., & Brisman, A. (2015). Climate change and peacemaking criminology: Ecophilosophy, peace and security in the "war on climate change". *Critical Criminology, 23*(4), 417–431.

McGrath, B. S. (2018). Countdown to day zero. *Time for Kids, 8*(18), 4–5.

McKenzie, D., & Swails, B. (2018). Day zero deferred, but Cape Town's water crisis is far from over. *CNN*. March 9. 2018. Retrieved 14 May, 2019, from https://www.cnn.com/2018/03/09/africa/cape-town-day-zero-crisis-intl/index.html

Meinen, D. (2017). The future (for) climate refugees: NGOs, security, and the politics of containment. *Canadian Network on Humanitarian History*, 23 January 2017. Retrieved 14 May, 2019, from http://aidhistory.ca/the-future-for-climate-refugees-ngos-security-and-the-politics-of-containment/

Molotch, S. (2012). *Against security: How we go wrong at airports, subways, and other sites of ambiguous danger*. Princeton: Princeton University Press.

Neocleous, M. (2008). *Critique of security*. Edinburgh: Edinburgh University Press.

Neocleous, M. (2014). *War power, police power*. Edinburgh: Edinburgh University Press.
Neocleous, M., Rigakos, G., & Wall, T. (2013). On pacification: Introduction to the special issue. *Socialist Studies/Études socialistes, 9*(2), 1–6.
Neocleous, M., & Rigakos, G. S. (Eds.). (2011). *Anti-security*. Ottawa: Red Quill Books.
O'Doherty, C. (2015). 120,000 Irish protest as the good boys of Europe turn bad. *Irish Central*, 22 November 2015. Retrieved 13 May, 2019, from https://www.irishcentral.com/opinion/cahirodoherty/120000-irish-protest-as-the-good-boys-of-europe-turn-bad
O'Halloran, M. (2016). Billing of water charges to be suspended next week. *The Irish Times*, 24 June 2016. Retrieved 13 May, 2019, from https://www.irishtimes.com/news/politics/oireachtas/billing-of-water-charges-to-be-suspended-next-week-1.2698891
O'Toole, F. (2014). The Irish rebellion over water. *The New York Times*, 19 December 2014. Retrieved 14 May, 2019, from https://www.nytimes.com/2014/12/20/opinion/fintan-otoole-the-irish-rebellion-over-water.html
Onishi, N., & Sengupta, S. (2018). In South Africa, facing 'day zero' with no water. *The New York Times*. January 31, 2019. Retrieved 14 May, 2019, from https://www.nytimes.com/2018/01/30/world/africa/cape-town-day-zero.html
Pérez-Peña, R. (2018). Cape Town pushes back 'day zero' as residents conserve water. *The New York Times*. February 20, 2018. Retrieved 14 May, 2019, from https://www.nytimes.com/2018/02/20/world/africa/cape-town-water-day-zero.html
Prosser, D. (2016) 'Six ways to profit by investing in water—the next big "super trend"'. *Money Observer*. Retrieved 14 May, 2019, from http://www.moneyobserver.com/our-analysis/six-ways-to-profit-investing-water-next-big-super-trend
Renshaw, J. (2017). East Coast refiner shuns Bakken delivery as Dakota Access Pipeline starts. *Reuters News Service*, 19 April 2017. Retrieved 13 May, 2019, from https://www.reuters.com/article/us-north-dakota-pipeline-pes/east-coast-refiner-shuns-bakken-delivery-as-dakota-access-pipeline-starts-idUSKBN17L0BJ
RTE News. (2016). EC to 'Evaluate' expert report on water charges. *RTE News Service*, 30 November 2016. Retrieved 13 May, 2019, from https://www.rte.ie/news/2016/1130/835595-european-commission-water/
Ryan, O. (2016). Sinn Féin unveils election billboard about water charges. *The Journal*, 21 January 2016. Retrieved 14 May, 2019, from http://www.thejournal.ie/sinn-fein-water-charges-3-2559660-Jan2016/
Samson, C., & Ravna, Ø. (2016). Civil liberties of indigenous people have long been suppressed at standing rock. Retrieved 24 December, 2016, from http://theconversation.com/civil-liberties-of-indigenous-people-have-long-been-suppressed-at-standing-rock-69817
Selby, J., Dahi, O., Fröhlich, C., & Hulme, M. (2017b). Climate change and the Syrian civil war revisited: A rejoinder. *Political Geography, 60*, 253–255.
Selby, J., Dahi, O. S., Fröhlich, C., & Hulme, M. (2017a). Climate change and the Syrian civil war revisited. *Political Geography, 60*, 232–244.
Sorisingtide. (2019). Retrieved 17 May, 2019, from http://sorisingtide.org/about/
South, N. (2012). Climate change, environmental (in)security, conflict and crime. In S. Farrall, D. French, & T. Ahmed (Eds.), *Climate change: Legal and criminological implications* (pp. 97–111). Oxford: Hart.
Stavriankas, A., & Selby, J. (2013). Militarism and international relations in the twenty-first century. In A. Stavriankas & J. Selby (Eds.), *Militarism and international relations: Political economy, security, theory* (pp. 3–18). New York: Routledge.
Stecker, T. (2016). Western Water: Jewell backs plan to remove 4 dams on Klamath. *E&E News*. October 18, 2916. Retrieved 24 December, 2016, from https://www.eenews.net/eenewspm/stories/1060044472.
Tertrais, B. (2011). The climate wars myth. *The Washington Quarterly, 34*(3), 17–29.
UN. (2013). *World population prospects: The 2012 revision*. New York: United Nations, Department of Economic and Social Affairs, Population Division.

Wall, T. (2013). Unmanning the police manhunt: Vertical security as pacification. *Socialist Studies/ Études socialistes, 9*(2), 32–56.

Welch, C. (2018). Why Cape Town is running out of water, and who's next. *National Geographic*. March 5, 2018. Retrieved 14 May, 2019, from https://news.nationalgeographic.com/2018/02/cape-town-running-out-of-water-drought-taps-shutoff-other-cities/

Welzer, H. (2012). Darfur: The first climate change war. *Audubon*, 24 July 2012. Retrieved 24 December, 2016, from http://www.audubon.org/news/darfur-first-climate-war

Zachos, E. (2017). How you can help fix the global water crisis. *National Geographic*. December 27, 2017. Retrieved 14 May, 2019, from https://news.nationalgeographic.com/2017/12/replenish-sandra-postel-water-conservation/

Chapter 3
Water Crimes Within Environmental Crimes

Lorenzo Segato, Walter Mattioli, and Nicola Capello

1 Introduction

Water stress is creating unprecedented tensions between freshwater users, with interconnected risks of various nature. This issue occurs when the demand for water exceeds the available amount during a certain period, or when poor quality restricts its use. Water stress causes deterioration of freshwater resources in terms of quantity (aquifer overexploitation, dry rivers, etc.) and quality (eutrophication, organic matter pollution, saline intrusion, etc.). The assumption concerning the secure supply of freshwater for the planet—even only for Europe—is wrong. Water covers more than 70% of the planet, but 97% of it is salted and found in the oceans, 2% is freshwater locked in snow and ice (e.g., glaciers), leaving less than 1% accessible for human use, distributed unequally. Only nine countries (Brazil, Russia, Canada, Indonesia, China, Colombia, the United States, Peru, India) share 60% of the total amount of freshwater (FAO, 2003; Miaschi, 2018), while 844 million people still lack even basic water services (United Nations, 2018). By the year 2025, 1.8 billion people will live in places and under conditions in which water is "scarce" (Arsenault, 2012; Ciampi, 2013; Johns, 2010). The global consumption of water is doubling every 20 years, more than twice the rate of human population growth; within the next three decades, demand for water from agriculture could increase by 50%, and for urban uses by between 50% and 70%; by 2035, the energy sector is projected to

L. Segato (✉) · N. Capello
Research and Action—REACT, Padova, Italy
e-mail: lorenzo.segato@re-act.it; nicola.capello@re-act.it

W. Mattioli
CREA—Research Centre for Forestry and Wood, Arezzo, Italy

Department for Innovation in Biological, Agro-food and Forest systems (DIBAF), University of Tuscia, Viterbo, Italy
e-mail: walter.mattioli@crea.gov.it; walter.mattioli@unitus.it

K. Eman et al. (eds.), *Water, Governance, and Crime Issues*,
https://doi.org/10.1007/978-3-030-44798-4_3

consume 85% more water (World Bank, 2013). These increased strains will create unprecedented conflicts between different water uses and interconnected risks between them (World Bank, 2016). In the period 2010–2018 (up to May 2018) 263 registered water related conflicts were reported (WWAP, 2019), p. 77 are currently ongoing, mainly in Asia and Africa (Factbook, 2019).

The progressive reduction in the quantity and quality of freshwater has major consequences for health, business success and security; harms and risks concerning water supplies are linked to multiple causes including economic growth and global warming, as well as crime. Reduced availability of freshwater, tensions over trans-boundary basins, water's strategic relevance for public safety and state security, widespread corruption and corporate interests, unequal distribution of resources, and large-scale investments in highly corrupt countries are the key drivers of criminal activities. Freshwater is becoming a "hot product" attractive for criminals (Clarke, 1999), potentially exposed to illicit interests of individuals, companies, terrorists, and organised crime. Both physical and logical water management infrastructures are exposed to cyber and traditional terrorist threats (ENISA, 2012).

Nevertheless, crimes against water remain difficult to assess, detect and prosecute, especially because legislation and policies on freshwater vary from country to country; hence, data are inconsistent or not comparable, as water crimes are often recorded under other offences—like fraud, corruption, theft, forgery/counterfeiting of documents, terrorism, or other crimes against health or the environment; moreover, academic and law enforcement interest in water crime issues (Bricknell, 2010; Brisman, McClanahan, & South, 2016; White, 2009) is very recent.

2 Environmental Crimes

2.1 Defining Environmental Crimes

A "crime" is considered by the International Classification of Crime for Statistical Purposes (ICCS) to be: "the punishable contravention or violation of the limits on human behaviour as imposed by national criminal legislation". Each criminal offence has a perpetrator—a person, corporation or institution—which is liable for the criminal behaviour in question. Environmental crime is a relatively new concept in international law (UNEP/INTERPOL, 2016), and an unequivocal and unanimously accepted definition of "environment" exists only in some countries (e.g., Italy) (Vagliasindi, Lucifora, & Bianco, 2015).

In general terms, "environmental crime" is often understood to describe illegal activities harming the environment and aimed at benefiting individuals or groups or companies through the exploitation of, damage to, trade or theft of natural resources, including serious crimes and transnational organised crime (UNEP/INTERPOL, 2016).

The problem with the definition of "environmental crime" arises from the fact that people use various synonyms such as "environmental criminality" or "criminality of environmental protection" (also "criminality of the environment", "green crimes", "crimes against the environment" and "ecocide") (Eman, Kuhar, & Meško, 2017). Specific reasons for such a situation depend on a set of reasons (Eman & Meško, 2014; Faure, 2009; Hannigan, 1995; INTERPOL, 2012; Meško, Dimitrijević, & Fields, 2011; Situ & Emmons, 2000; UNEP/INTERPOL, 2016), namely:

1. environmental crime is related to technical development and progress, which is why new forms of environmental crime are emerging continuously;
2. environmental crime is very diverse all around the world (it is inherent to country, economic system, environmental and biological systems, etc.);
3. environmental crime is a complex phenomenon, which is why its investigation is usually lengthy and complicated;
4. environmental crime is difficult to detect (no significant visibility of the offender);
5. the collectivity and anonymity of victims lead to a rather abstract and vague perception of damage;
6. offenders of environmental crime are often linked to other parts of the world, which means that foreign legal standards should be considered in addition to domestic ones;
7. environmental criminal offences are typically characterized by their frequency, irresistibility, poor measurement and social apathy, which further complicates their detection and punishment;
8. certain environmental criminal offences tend to be committed during the performance of a profession or economic activity and abuse of trust (the acquisition of financial gain through environmental white-collar crime);
9. boundaries between legal and illegal acts are often vague in the field of ecology (circumvention of environmental threats to the legal order of a country);
10. environmental justice tackles different issues in different countries (e.g. environmental racism, organised crime groups involved in thefts of natural resources, etc.);
11. international environmental law is an imperfect system for the protection of the environment since it is sometimes excessively broad and vague or depending on national interests elsewhere.

Environmental crime typically refers to any breach of a national or international environmental law or convention that exists to ensure the conservation and sustainability of the world's environment (Elliott, 2007; Waleij, Liljedahl, Mattsson, & Simonsson, 2013). For Clifford (1998), environmental crime is an act committed with the intention of damaging or causing damage to the ecological and biological systems in order to ensure a business or personal benefit. Situ and Emmons (2000) define environmental crime as necessitating the creation of environmental legislation, because any behaviour that does not violate the law, however threatening or aggressive, is not a criminal act. Dini (2007) defines environmental crime as an act involving the already illegal production and/or distribution of goods and services (e.g., illegal trade in wildlife, trafficking of ozone-depleting substances, illegal transport, and trade of hazardous or radioactive waste, illegal fishing and poaching, timber logging, etc.).

Although there is no universally agreed definition of environmental crime, this collective term is often understood to describe illegal activities harming the environment and aimed at benefiting individuals or groups or companies through the exploitation of, damage to, trade or theft of natural resources, including serious crimes and transnational organized crime (UNEP/INTERPOL, 2016).

2.2 *Types and Scale of Environmental Crimes*

For a long time, five areas have been considered of major importance in relation to environmental crime: illegal trade in wildlife; illegal logging and its associated timber trade; illegal, unreported and unregulated fishing; illegal trade in controlled chemicals (including ozone-depleting substances) and illegal disposal of hazardous waste (INTERPOL, 2012); references to "water" were missing.

More recently, new types of environmental crime have been taken into account, no longer restricted to iconic wildlife and rare wood species alone: a recent report of UNEP/INTERPOL (2016) points out that the term "environmental crime" covers not only illegal trade in wildlife, but also forestry, illegal logging and fishery crimes, illegal dumping of waste including chemicals, smuggling of ozone-depleting substances, surface water pollution and illegal mining (Fig. 3.1).

A broader understanding of environmental crime includes threat finance from exploitation of natural resources such as minerals, oil, timber, charcoal, marine resources; financial crimes in natural resources; laundering, tax fraud and illegal trade in hazardous waste and chemicals; as well as the environmental impact of illegal exploitation and extraction of natural resources. Second only to trade in drugs and weapons, environmental crime is the largest illegal business in the world (UNEP/INTERPOL, 2016). INTERPOL (2016) estimates that global wildlife crime is worth billions of dollars a year; the World Bank (2016) states that illegal logging costs developing countries $15 billion in lost revenue and taxes. In the mid-1990s, around 38,000 tons of CFCs (chlorofluorocarbons) were traded illegally every year—equivalent to 20% of global trade in CFCs and worth $500 million; and in 2006 up to 14,000 tons of CFCs were smuggled into developing countries (EIA, 2008).

2.3 *Causes and Consequences of Environmental Crimes*

The root causes of environmental crimes vary greatly: these are primarily the low risks and high profits in an environment made permissive by poor governance and widespread corruption, minimal budget allocated to police, prosecution and courts, inadequate institutional support, political interference and low employee morale, as well as minimal benefits to local communities. The situation is especially critical

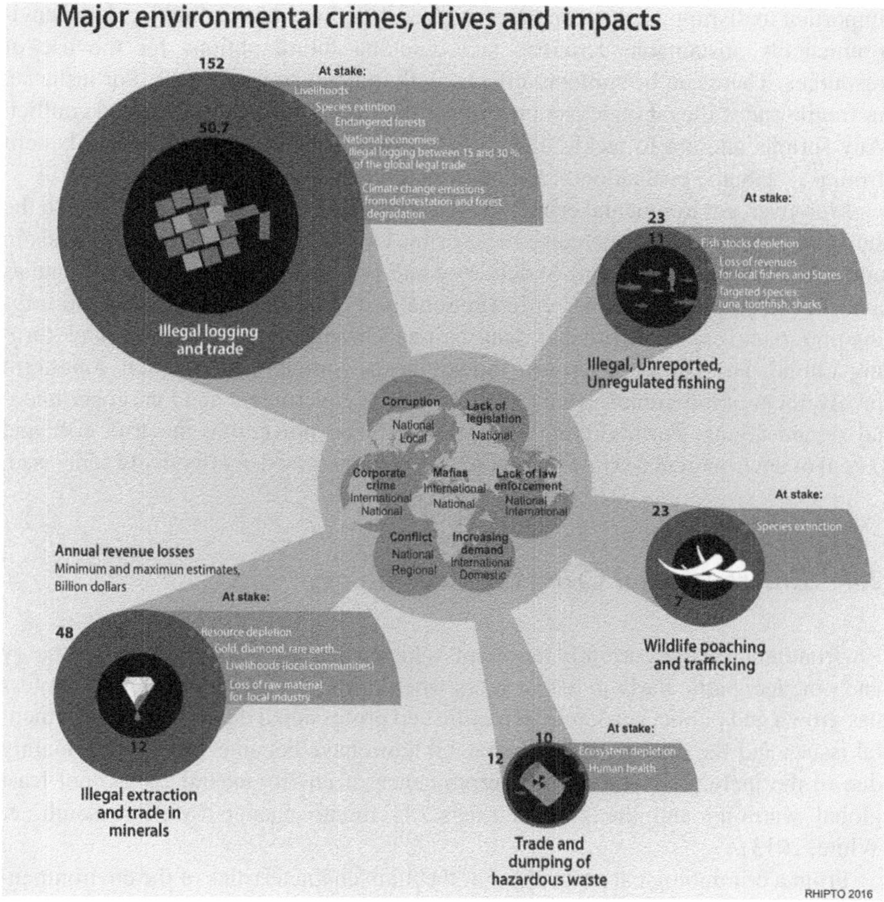

Fig. 3.1 Major environmental crimes, drives and impacts (UNEP/INTERPOL, 2016)

when it comes to support to prosecution and courts/the judiciary in many developing countries.

Unlike any other known crime, environmental crimes are aggravated by their impact on the environment and their costs for future generations. Deforestation, dumping of chemicals and illegal fisheries cause loss of ecosystem services such as clean air and clean water, extreme weather mitigation, food security, and even health and wellbeing. They also deprive governments of much-needed revenues and undermine legal businesses. An additional by-product of environmental crime is that it undermines peace. It is not surprising that the UN Security Council has recognised the serious threat to security posed by environmental crime, with UN reports pointing to armed groups and potentially even terrorists sustained through the spoils of this rising criminal industry.

Some of the consequences of environmental crime are irreversible. The economic, environmental and health impacts of illegal trade can be sufficiently important to disrupt whole economies and ecosystems, undermining legal and environmentally sustainable activities and reducing future options for the use of resources. There can be spillover effects, with indirect consequences. For instance, in fragile states illegal trade can undermine the rule of law and fuel armed conflict. Any serious attempt to tackle illegal trade in environmental goods would benefit from a systematic evaluation of the impacts (OECD, 2012).

Moreover, environmental crimes are a catalyst for corruption (EIA, 2008). In the same way that criminals perceive these crimes as an easy option, so individuals in corporate or official positions of authority and power view environmental crime as a chance to cash in. Examples of institutionalised corruption are: signing and forging import and export certificates; facilitating the transport of illicit goods and 'turning a blind eye'. Far more serious, and yet just as common, is complicit, long-term involvement of individuals from the police, army, government and intergovernmental organisations. Furthermore, corruption may be preventing the true cost and extent of environmental crime from being properly assessed or effectively addressed.

2.4 Environmental Crimes in Criminology

Environmental crime currently represents a big challenge for criminal justice theory and practice, particularly in recent years when environmental protection awareness has grown and public, academic, scientific and professional debates on environmental issues and the causes of environmental harm have become widespread, mainly due to the increasing urgency and prominence of environmental issues, not least global warming and the gradual crisis of climate change (Meško, South, & White, 2013).

From a criminological point of view, the main characteristics of the environmental crimes are (Eman, Meško, & Fields, 2009):

- the anonymity of the victim;
- the insignificant visibility of the perpetrator of such acts;
- an abstract comprehension of the damage caused to the environment;
- a widespread imperceptibility of ecological crimes and recognised complications involved with generating evidence with which to prosecute these crimes.

Environmental harm includes legal or illegal acts or omissions against the environment committed by man (Heckenberg, 2009). The link between environmental issues and criminology is captured by so-called 'green criminology', a new discipline that has arisen from advances and concerns from outside the field as such.

Green criminology developed from discussions about environmental crime and evolved into a science about environmental harm (White, 2009). In other words, green criminology examines deviance against the environment and violations of environmental protection legislation (Eman & Meško, 2014) and addresses forms of crime that harm the environment but are often ignored in criminological research (Lynch & Stretsky, 2007). This relatively new area of research is concerned with stretching the boundaries of mainstream criminology to accommodate issues of global significance, while also utilising the insights of conventional criminology to illuminate ways in which to understand and to respond to environmental harm (Eman, Meško, Dobovšek, & Sotlar, 2013).

The emergence of green criminology dates to the late 1960s and the beginning of the 1970s, when environmental sociologists and critical criminologists first emphasised the problem of environmental pollution and destruction (Eman et al., 2013). Throughout history, three directions of green criminology developed (radical-critical, sociological-theoretical, and sociological-philosophical) (Lynch & Stretsky, 2003) which later converged into what is today known as green criminology, a term proposed by Lynch (1990) and (Koser Wilson, 1999).

Finally, green criminology can be described as a discipline that adopts multidisciplinary and interdisciplinary approaches to research environmental crime, environmental harm, environmental legislation, environmental regulations, environmental protection measures and public responses to violations. For these reasons, green criminology is well equipped to address environmental crimes in the water sector.

3 Defining Water Crimes

In global terms, crime against water is considered as an emerging global issue (INTERPOL, 2012) and water management as an emerging environmental crime (INTERPOL, 2016). These topics are becoming ever more relevant as nation-states and enforcement agencies increasingly realise the scope and nature of the problem (INTERPOL, 2016); however, intelligence is scarce, and information fragmented. The main issue is the absence of a common definition of water crimes, water-related crimes and water security. Moreover, water crime does not enjoy a specific status among environmental crimes; instead, there are only some general observations: water may be the environmental resource damaged by a crime (e.g. surface water pollution, or fraudulent water quality reporting); the object of a crime (e.g. drinking water theft or corruption on the part of private companies related to the economic control/exploitation of water); or the means of the crime (e.g. intentional flooding or the deliberate poisoning of a water supply).

Water-related crimes are often recorded under other offences—like fraud, corruption, trafficking, falsification of documents, terrorism—due to the absence of a systematic analytical approach. For instance, within the International Classification of Crime for Statistical Purposes (UNODC, 2016) water and water-related crimes might fall into the categories of acts against the natural environment, acts leading to death or intending to cause death, acts leading to harm or intending to cause harm to the person, involving fraud, deception or corruption, acts against public safety and state security, acts against property only.

The adoption of a universal working definition of water crimes must include all meaningful forms of offences against water bodies, based on—but not necessarily limited to - the provisions contained in the criminal law of different countries.

The European Union defines freshwater as "naturally occurring water having a low concentration of salts, which is often acceptable as suitable for abstraction and treatment to produce drinking water" (art. 2b Dir 91/676/EEC). Surface water, groundwater, and water services are defined in the EC Water Framework Directive 2000/60/EC and the Drinking Water Directive 98/83/EC. The EU Directive 2000/60/EC (art. 1) establishes a framework for the protection of inland surface waters, transitional waters, coastal waters and groundwater, while within Art. 2 "surface water" is taken to indicate: inland waters (except groundwater) and transitional waters and coastal waters (except in regard to chemical status, for which it shall also include territorial waters); the term "groundwater" indicates all water which is below the surface of the ground in the saturation zone and direct contact with the ground or subsoil; the term "water intended for human consumption" has the same meaning as under Directive 80/778/EEC, which excludes natural mineral waters and medicinal waters, as amended by Directive 98/83/EC, where art. 1 concerns the quality of water intended for human consumption. Lastly, the term "water services" indicates all services which provide water for households, public institutions, or any economic activity, including: (a) abstraction, impoundment, storage, treatment and distribution of surface water or groundwater; (b) waste-water collection and treatment facilities which subsequently discharge into surface water.

The Water Crimes Project (2016) adopted the following definitions:

- *Water crime* is defined as any punishable contravention or violation of the limits on human behaviour as imposed by national criminal legislation, against surface, and groundwater, or against water services. This also includes any intentional act that can potentially harm or damage water;
- *Water-related crime* is defined as any punishable contravention or violation of the limits on human behaviour as imposed by national criminal legislation, which uses surface and groundwater, or water services, as a means for committing other crimes.

3.1 Types of Water Crimes

Water crimes range from pilfering of water from pipelines to illegal waste management by a factory, up to pollution or terrorist threats. Water theft includes the illegal acquisition of natural water through pumping, impoundment, or diversion of water from irrigation channels, river systems, dams, or groundwater courses, damaging or removing water meters and physically connecting to water distribution pipes as well as stealing aid tanks (The Global Initiative, 2014). Water crimes may affect freshwater through acts that undermine the quality, quantity, accessibility or availability of water. Moreover, water crimes go far beyond being a 'simple crime', because they are linked to convergence crimes, as it happens for the other environmental crimes.

A clear example of this is the correlation between corruption and access to improved drinking water in Sub-Saharan Africa; The more corrupt a country is, the smaller the fraction of its population that has access to improved drinking water (Fig. 3.2).

Moreover, a recent phenomenon is the smuggling of water through complex network chains and water mafias: ways of smuggling vary greatly and include, among others, the development of illegal pipelines, illegal truck deliveries, as well as the co-optation of water regulators complicit in licensing fraud and broader government acquiescence to illegal water delivery. Illegally sourced and smuggled water is used for personal consumption, agriculture, industry, and sometimes for other illegal activities, such as the production of illegal narcotics. Other examples of different types of water crimes identified by the European Union Council (2016) include: river and marine pollution, manipulation of sampling methods to avoid treatment costs, fraud and illegal trafficking of water, terrorism and cyber threats to water management operations, illegal waste discharges from factories, and unauthorized consumption from the water network.

Water crime has been tentatively classified (Segato, Mattioli, Capello, & Migliorini, 2017; Water Crimes Project, 2016) in the following types:

1. *Water corruption*: to be distinguished into a) grand corruption, which involves political decision-makers, exploitation of natural resources, international cooperation, and large-scale investments and large procurement contracts; and b) petty corruption, which people or companies may experience daily when they are requested to pay money in order to access water services or to avoid inspections and fines;

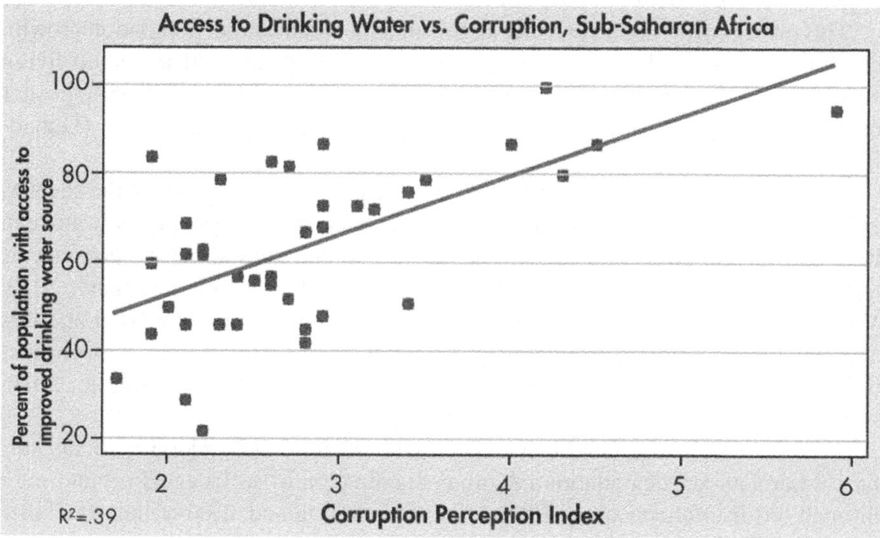

Fig. 3.2 Access to drinking water vs. corruption, Sub-Saharan Africa (Stålgren, 2006)

2. *Water organized crime*: the activity of criminal organizations that have taken control over the management of water or water services within a given territory;
3. *Water pollution*: offences against the natural environmental and, especially, directly against water quality;
4. *Water theft*: reduction of the water quantity, carried out by users of the water supply system in order to provide an economic advantage through physical alteration of the water supply system;
5. *Water fraud*: consisting in fiscal artifices aimed at adulterating the measure of water quality and/or quantity in order to reach an illicit gain;
6. *Water terrorism*: threat of terrorist attacks targeting water sectors, particularly water quality (e.g., poisoning) or availability (e.g., attack to critical infrastructures). It might also include taking control over water services for the illicit financing of terrorist activities or for retaliation;
7. *Water cyber-attacks*: intrusion into the ICT system, manipulation of information or networks, data destruction, etc. of water management companies. It includes ransomware and malware attacks.

3.2 Causes and Consequences of Water Crimes

Water security is weakened by gaps and loopholes in legislation, and exposed to criminal activities by individuals, companies, and criminal groups. Over time, the "blue gold" (Barlow, 2001) has become an opportunity and a product for criminals, driven by perceptions of low risk and high profit, attracting the greater interest of organised crime groups (UNDP, 2011).

This emerging problem is further aggravated by urbanisation, population growth, environmental degradation, water pollution, climate change and increased living standards. As surface water depletes, there is increased pressure for groundwater usage, which is more difficult to regulate for use, abuse, and theft (Felbab-Brown, 2015).

The cumulative impact of water theft has the potential to jeopardise the sustainability of water resources. For example, over-extraction of coastal groundwater can have detrimental effects for terrestrial, freshwater, and coastal marine ecosystems that rely on water flows through creeks and rivers (Poff & Zimmermann, 2010; Vörösmarty et al., 2010). It can also cause seawater intrusion into coastal aquifers and render these aquifers unsuitable for irrigation, therefore impacting land productivity and farmers' viability (Greiner, Fernandes, McCartney, & Durante, 2016; Narayan, Schleeberger, & Bristol, 2007; Werner, 2010).

Water contamination can be associated with depletion of underground and surface water sources, degradation of soil, contamination of surface and groundwater through fertiliser/chemical or effluent runoff from farmland, destruction of wildlife habitat, and reduction of biodiversity.

Recent weather events all around the planet have given rise to enhanced concerns about the impact and consequences of toxic chemicals and hazardous waste materials being swept up in flood waters and polluting freshwater systems. The storage of legacy mining waste, stockpiling of tires, repositories for chemical wastes and so on are indirectly implicated in water-related crime as far as these activities may violate relevant regulatory regimes and thus contribute to greater harm than might normally be expected.

3.3 *Risks and Threats of Water Crimes*

While awareness of environmental crimes is gaining momentum at the international level, crimes against water are an emerging issue of global concern, as it has major ramifications for water use, ecosystem health and functioning, and for the biodiversity that underpins our ecosystems.

Growing awareness about water crimes is mainly related to the fact that water requirements to sustain ecosystem health and biodiversity in open surface waters and wetlands must be well aligned with human needs and the delivery of a range of ecosystem goods and services to society (Bernhardt, Bunn, Hart, & Van Wilgen, 2006; Postel & Richter, 2003).

Water security is among the top global risks in terms of development impact (WEC, 2016) but remains a challenge for many countries that today are coping with complex water issues that cut across economic sectors (World Bank, 2016).

Corruption also plays a significant role in water smuggling: petty bribes to government officials and water suppliers are used to falsify meter readings, avoiding disconnections, and concealing illegal connections. The rising demand for water only serves to reinforce these corrupt practices and enforce a sense of discretionary power among officials, consequently weakening a country's rule of law (The Global Initiative, 2014).

Some causes of corruption in the water sector are (UNDP, 2011):

- weak enabling environment (e.g. the weak rule of law, political environment, etc.);
- low levels of accountability and transparency:
- weak technical and management capacity;
- political capture of water policies and projects:
- gaps in laws and policies;
- the international dimension of corruption (multi-national companies, etc.).

Whereas the scope of corruption varies substantially across the sector and between different countries and governance systems, estimates by the World Bank (2016) suggest that 20–40% of water sector finances are being lost to dishonest and corrupt practices.

Furthermore, both physical and logical water management infrastructures are exposed to cyber and traditional terrorist threat (ENISA, 2012). Recent terrorist events around the globe highlight that the possibilities of intentional poisoning of freshwater supply also need to be considered carefully and seriously as part of the water crimes. Assessing these threats involves not only an analysis of broad counter-terrorism strategies, but also an examination of the geographical, social and economic contexts within which water supply (its quantity and quality) is managed and monitored.

Finally, the potential threat of water crimes to national and international security, financial wellbeing, and human health in a period witnessing rapid economic, social and environmental change should be duly considered.

4 Conclusions

Water is one of the most basic human needs and is indispensable to almost all economic activities, including agriculture, energy production, industry, and mining (World Bank, 2013). Nowadays, water is under unprecedented pressure as growing populations and competing economic sectors demand more of it leaving insufficient water to meet human needs, as well as sustaining the environmental flows that keep our ecosystems healthy. Crime is emerging as one of the future threats to freshwater availability.

Groundwater reserves are depleting in many places, leaving current and future generations with close to no buffer against increased climate variability, and, without effective regulation and suppression of water crimes, the sustainability, long-term viability, and inclusive and equitable use of water cannot easily be achieved (Felbab-Brown, 2015). To date, awareness, intelligence and studies on water security appear mainly focused on critical infrastructures, and more developed in other regions of the world than in Europe, where this type of environmental crimes has never been systematically analysed. Therefore, data on water crimes are largely missing. This gap in knowledge on criminal phenomena hampers the capacity of institutions and law enforcement agencies to design effective strategies for preventing these crimes and protecting water from being attacked or damaged. Consequences of water crimes—such as the contamination of a water source, or the interruption of water service for ransom—can be hardly mitigated, as the supply of freshwater must be guaranteed regardless the costs of service. This is relevant also for water service companies. Therefore, criminological research shall focus on crime patterns, collecting data from European countries, and on risk assessment analysis of these emerging crimes.

The situation is expected to worsen, as water demand (and price) is expected to increase by 55% worldwide between 2000 and 2050, making water very attractive for criminals and criminal organisations. In fighting environmental crimes, a strong regulatory regime and effective prevention mechanisms, including anti-corruption measures, may be just as important as criminal law tools. Environmental crime is a haven for corruption at all levels and, unless corrupt officials are tackled, efforts to combat environmental crime will be impeded, a fact that should be acknowledged within cross-cutting resolutions on environmental crime within the United Nations and within the UN Convention against Corruption (EIA, 2008). This is particularly true when it comes to crimes against water.

International criminal law conventions (or a single convention) could be conceived to have comparable obligations of criminalisation and an extension of criminal jurisdictions ensuring that no safe havens remain for the offenders (UNODC, 2016).

References

Arsenault, C. (2012). *Risk of water wars rises with scarcity*. Retrieved 17 May, 2019, from http://www.aljazeera.com/indepth/features/2011/06/2011622193147231653.html

Barlow, M. (2001). Blue gold. In *The global water crisis and the commodification of the world's water supply*. Ottawa: National Chairperson, Council of Canadians.

Bernhardt, E. S., Bunn, S. E., Hart, D. D., & Van Wilgen, B. (2006). Perspective: The challenge of ecologically sustainable water management. *Water Policy, 8*, 475–479.

Bricknell, S. (2010). Environmental crime in Australia. In *Research and public policy series on water theft* (pp. 103–114). Canberra: Australian Institute of Criminology.

Brisman, A., McClanahan, W., & South, N. (2016). Water security, crime and conflict. Criminology and criminal justice, international and comparative criminology. In *Oxford handbook*. Oxford: Oxford University Press.

Ciampi, M. (2013). 'Water Divide' in the Global Risk Society. *International Review of Sociology, 23*(1), 243–260.

Clarke, R. V. (1999). Hot products: understanding, anticipating and reducing demand for stolen goods. In *Policing & reducing crime: Police research series* (p. 112). London: RDS.

Clifford, M. (1998). *Environmental crime—Defining environmental crime*. Retrieved 17 May, 2019, from http://law.jrank.org/pages/11964/Environmental-Crime-Defining-environmental-crime.html

Dini, J. W. (2007). International environmental crime. *Plating and Surface Fishing, 94*(7), 58–59.

Elliott, L. (2007). Transnational environmental crime in the Asia-Pacific: An "un(der)securitized" security problem? *The Pacific Review, 20*(4), 499–522.

Eman, K., Kuhar, S., & Meško, G. (2017). Water crimes and policing. *Varstvoslovje, 19*(2), 105–119.

Eman, K., & Meško, G. (2014). Environmental crime in criminology. In *Crime phenomena and development of a green criminology in Slovenia*. Saarbrücken: Scholar's Press.

Eman, K., Meško, G., Dobovšek, B., & Sotlar, A. (2013). Environmental crime and green criminology in South Eastern Europe—Practice and research. *Crime, Law, & Social Change, 59*, 341–358.

Eman, K., Meško, G., & Fields, C. B. (2009). Crimes against the environment: Green criminology and research challenges in Slovenia. *Varstvoslovje, 11*(4), 574–592.

Environmental Investigation Agency (EIA). (2008). *Environmental crime. A threat to our future*. London: EIA. https://eia-international.org/report/environmental-crime-2/

European Union Agency for Network and Information Security (ENISA). (2012). *NCSS good practice guide. Designing and implementing National Cyber Security Strategies*. Heraklion: ENISA.

European Union Council. (2016). *Council conclusions on countering environmental crime*. Retrieved 17 May, 2019, from http://data.consilium.europa.eu/doc/document/ST-15412-2016-INIT/en/pdf

Factbook. (2019). *Environment, conflict and cooperation*. Retrieved 17 May, 2019, from https://factbook.ecc-platform.org/conflicts?f[]=index_type_of_resource:5&f[]=conflict_active_years:2019&f%5B%5D=index_type_of_resource%3A5
FAO. (2003). *Review of world water resources by country*. Retrieved 26 May, 2019, from http://www.fao.org/3/Y4473E/y4473e00.htm#Contents
Faure, N. (2009). *Environmental crimes*. Maastricht: Maastricht University, Faculty of Law.
Felbab-Brown, V. (2015). *Water crimes: A global crisis on the rise*. Retrieved 26 May, 2019, from https://www.brookings.edu/on-the-record/water-crimes-a-global-crisis-on-the-rise/
Greiner, R., Fernandes, L., McCartney, F., & Durante, J. (2016). Reasons why some irrigation water users fail to comply with water use regulations: a case study from Queensland. *Land Use Policy, 51*, 26–40.
Hannigan, A. J. (1995). *Environmental sociology. A social construction perspective*. London, New York: Routledge.
Heckenberg, D. (2009). Studying environmental crime: Key words, acronyms and sources of information. In R. D. White (Ed.), *Environmental crime: A reader* (pp. 9–24). Cullompton: Willan Publishing.
INTERPOL. (2012). *Environmental crime. International Criminal Police Organization*. Retrieved 17 May, 2019, from http://www.INTERPOL.int/Crime-areas/Environmental-crime/Environmental-crime
INTERPOL. (2016). *Strategic report. Environment, peace and security. A convergence of threats*. Retrieved 17 May, 2019, from https://europa.eu/capacity4dev/unep/document/strategic-report-environment-peace-and-security-convergence-threats
Johns, C. (2010). Editor's Note. In special issue: "Water: our thirsty world". *National Geographic, 217*(4), 4.
Koser Wilson, N. (1999). Eco-critical criminology—An introduction. *Criminal Justice Policy Review, 10*(2), 155–160.
Lynch, J. M. (1990). The greening of criminology: a perspective for the 1990's. *The Critical Criminologist, 7*(2), 11–12.
Lynch, J. M., & Stretsky, B. P. (2003). The meaning of green: Contrasting criminological perspectives. *Theoretical Criminology, 7*(2), 217–238.
Lynch, J. M., & Stretsky, B. P. (2007). Green criminology in the United States. In P. Beirne & N. South (Eds.), *Issues in green criminology: Confronting harms against environments, humanity and other animals* (pp. 217–238). Portland: Willan Publishing.
Meško, G., Dimitrijević, D., & Fields, C. B. (2011). *Understanding and managing threats to the environment in South Eastern Europe*. Dordrecht: Springer.
Meško, G., South, N., & White, R. (2013). Editors' introduction. *Crime, Law, & Social Change, 59*, 249–250.
Miaschi, J. (2018). Which country has the most fresh water? *World Atlas*. http://worldatlas.com/articles/countries-with-the-most-freshwater-resources.html
Narayan, K., Schleeberger, C., & Bristol, K. (2007). Modelling seawater intrusion in the Burdekin Delta irrigation area, North Queensland, Australia. *Agriculture Water Management, 89*, 217–228.
OECD. (2012). Illegal trade in environmentally sensitive goods. In *OECD trade policy studies*. Paris: OECD Publishing. Retrieved 26 May, 2019, from https://doi.org/10.1787/9789264174238-en
Poff, N., & Zimmermann, J. (2010). Ecological responses to altered flow regimes: a literature review to inform the science and management of environmental flows. *Freshwater Biology, 55*, 194–205.
Postel, S., & Richter, B. (2003). *Rivers for life: managing water for people and nature*. Washington, D.C.: Island Press.
Segato, L., Mattioli, W., Capello, N., & Migliorini, M. (2017). Environmental crimes in the water sector. *European Water, 60*, 373–379.
Situ, Y., & Emmons, D. (2000). *Environmental crime: The criminal justice system's role in protecting environment*. Thousand Oaks: Sage Publications.

Stålgren, P. (2006). Corruption in the water sector: Causes, consequences and potential reform. *Swedish Water House Policy Brief* Nr. 4. SIWI, 2006.

The Global Initiative. (2014). *Water, water everywhere: charting the growth of organized water theft*. Retrieved 17 May, 2019, from https://globalinitiative.net/water-smuggling/

UNEP/INTERPOL. (2016). The rise of environmental crime. In *A growing threat to natural resources, peace, development and security. A UNEP/INTERPOL Rapid Response Assessment*. United Nations Environment Programme and RHIPTO Rapid Response-Norwegian Center for Global Analyses.

United Nations (UN). (2018). *Sustainable development goal 6 synthesis report 2018 on water and sanitation (executive summary)*. Retrieved 26 May, 2019, from https://www.unwater.org/publications/executive-summary-sdg-6-synthesis-report-2018-on-water-and-sanitation/

United Nations Development Program (UNDP). (2011). Fighting corruption in the water sector. In *Methods, tools and good practices*. New York: UNDP.

United Nations Office on Drugs and Crime (UNODC). (2016). *The International Classification of Crime for Statistical Purposes (ICCS)*. Vienna: United Nations Office on Drugs and Crime. Retrieved 17 May, 2019, from http://www.unodc.org/documents/data-and-analysis/statistics/crime/ICCS/ICCS_English_2016_web.pdf

Vagliasindi, G. M., Lucifora, A., & Bianco, F. (2015). *Fighting environmental crime in Italy: A country report. Study in the framework of the EFFACE research project*. Catania: University of Catania.

Vörösmarty, C., McIntyre, P., Gessner, M., Dudgeon, D., Prusevich, A., Green, P., et al. (2010). Global threats to human water security and river biodiversity. *Nature, 467*, 555–561.

Waleij, A., Liljedahl, B., Mattsson, K. M., & Simonsson, L. (2013). The role of environmental crime in terrorism, conflict and criminality. In M. T. Lindell, L. M. Hellström, & Å. Wiss (Eds.), *Strategic outlook 2013*. Stockholm: Swedish Defence Research Agency (FOI).

Water Crimes Project. (2016). *Inventory, outlook and assessment of emerging environmental crimes against water in Europe*. Retrieved 12 May, 2019, from http://www.watercrimes.eu/

Werner, A. (2010). A review of seawater intrusion and its management in Australia. *Hydrogeological Journal, 18*, 281–285.

White, R. D. (Ed.). (2009). *Environmental crime: A reader*. Cullompton: Willan Publishing.

World Bank. (2013). *Water resources management: sector results profile*. Retrieved 17 May, 2019, from http://www.worldbank.org/en/results/2013/04/15/water-resources-management-results-profile

World Bank. (2016). High and dry. Climate change, water, and the economy. *Water Global Practice*. Retrieved 17 May, 2019, from https://openknowledge.worldbank.org/handle/10986/23665

World Economic Forum (WEC). (2016). *The global risk report* 2016. Retrieved 17 May, 2019, from https://www.weforum.org/reports/the-global-risks-report-2016

WWAP (UNESCO World Water Assessment Programme). (2019). *The United Nations world water development report 2019: Leaving no one behind*. Paris: UNESCO.

Chapter 4
Water and Organised Crime

Katja Eman and Rob White

1 Introduction

Water is a basic requirement of human life. It is always needed and, therefore, marketable. Today, water is one of the most valuable resources in the world, even more so than gold or oil. In the late 1990s, the re-conceptualization of water as a tradeable commodity occurred (Alston & Mason, 2008), accompanied by privatisation of its ownership and management, which caused numerous changes concerning the rights of the users and the ways of its use. Scarce supply and unequal access to water have also been affected by factors such as climate change and excessive agricultural irrigation (Johnson, South, & Walters, 2015; UNESCO, 2012). Restricted quantities of clean water make it a particularly valuable property for those who own and control it. It also opens the door to corruption and the involvement in organised criminal groups in its supply and distribution.

Environmental safety and wellbeing—including provision of clean drinking water—is closely linked to national and international security. The main pressure on governments in protecting the environment comes from the private business sector, which inherently strives for the maximization of profit even if this is at the expense of the environment (Bisschop, 2011; Elliott, 2009; Klenovšek & Meško, 2011). When environmental protection laws are passed, many companies will comply; but others will try to circumvent the norms that are in place to protect the environment. This can translate into situations where further pollution or depletion of natural resources occurs with the assistance of organised criminal groups and/or illegal services which enable companies and individuals to continue their previous

K. Eman (✉)
Faculty of Criminal Justice and Security, University of Maribor, Ljubljana, Slovenia
e-mail: katja.eman@fvv.uni-mb.si

R. White
School of Social Sciences, University of Tasmania, Hobart, TAS, Australia
e-mail: R.D.White@utas.edu.au

K. Eman et al. (eds.), *Water, Governance, and Crime Issues*,
https://doi.org/10.1007/978-3-030-44798-4_4

profitable activities. An example of this is waste management and the contracting out of waste removal by legitimate companies to unscrupulous operators who dispose of the waste illegally (Dobovšek, 1997; Massari & Monzini, 2004; Situ & Emmons, 2000; Watson, 2005).

Longer periods of water shortages, together with the competing rights of irrigators versus environmentalists and urban versus rural populations raised several huge challenges for politics and policymakers (Alston & Mason, 2008, p. 214). The most important is the fact that every human has a right to water and sanitation. The fact that water was accepted as a tradeable commodity and that people need it for their survival, offered an opportunity for never-ending business. Thus, the increasing significance given to the commodification of water and the decreasing significance of the environmental uses of water opened the door to the water trade as big business. The prioritization of economic dimensions of water began in the 1990s and has accelerated since then.

Freshwater became a lucrative investment because its scarcity means a huge profit for water-related businesses. The rising freshwater industry is estimated to be worth around US$1 trillion a year (Johnson et al., 2015, p. 149). Gleick (1998, p. 571) notes that the human right to water, the limits of fresh water sources, and the need for sustainable development were pushed into the background as secondary considerations because of the logic of privatization and prevailing economic models. Water ownership and use have been reconstructed in law and policy in the interests of private companies. This situation creates inequalities and injustice in the distribution of freshwater. Existing laws are still subject to violation in the form of water-related corruption and violence, water theft, water market price fixing, and violations of the regulations on water quality. Here corruption also plays a significant role, as transparency in the water sector is often limited (Eman, Furdi, Hacin, & Dobovšek, 2016; Spapens, White, van Uhm, & Huisman, 2018). Moreover, the possibilities for making huge profits always attract organized crime.

2 Organized Environmental Crime

Much has been written about how organized criminal groups are involved in environmental crime activities (Newman, 2001; Siegel, 2001; Hayman & Brack, 2002; Massari & Monzini, 2004; Schmidt, 2004; Watson, 2005; Lin, 2005; Elliott, 2007, Elliott, 2009; Tacconi, 2008; Stølsvik, 2010; Eman, 2010; Bisschop, 2011; Klenovšek & Meško, 2011; Nurse, 2015; White, 2016; van Uhm, 2016). Less attention has been given to water crimes, but this is changing (Eman, Kuhar, & Meško, 2017; Moretti et al., 2017; Sergi & South, 2016; Spapens et al., 2018; The Global Initiative, 2014).

Environmental crime has become attractive to organized criminal groups mainly because of the possibility of large earnings and the relative lack of environmental protection including adequate criminal legislation. The United Nations Convention Against Transnational Organised Crime (2000, Article 2) defines an organized criminal group as "a structured group of three or more people over a longer period of time and acting in concert to commit one or more serious offenses under this Convention, directly or indirectly, for a financial or other material benefit." This definition is very broad as are other definitions that emphasise hierarchy and the presence of a formalised structure in pursuit of money through illegal activities (Abadinsky, 2016; FBI, 2010). In these representations, organised criminal groups maintain their position by threats, violence, bribery of state officials, and extortion, and have a strong impact on the people of the local community, region or country as a whole (Federal Bureau of Investigation (FBI), 2010). Organized crime of this nature is a crime of structured groups, which usually includes offering illegal goods and services. What is a more, organized crime of this type and scale establishes a parallel state system and undermines the state monopoly over the means of coercion and the ability to levy taxes pose and thus presents a direct threat to the state and social structures in the places where it operates (Dobovšek, 2009, p. 18).

However, the criminal networks involved in environmental and related cross-over crimes are many and varied (Ayling, 2013). Environmental crime is a high-profit and low-risk activity. Its major drivers are the demand for environmental commodities and the lack of effective deterrents (Interpol/UNEP Environment, 2016). For organized criminal groups, environmental crime is extremely lucrative but with comparatively low risk due to several situational factors on local, national and global levels that equally contribute to environmental crime (Interpol/UNEP Environment, 2016). At the local level, environmental crime is an alternative to poverty for disadvantaged populations. It can provide an additional, and sometimes the only, source of earning for them. Criminal networks often exploit the needs of vulnerable communities, i.e. tasking locals with poaching, logging, or mining in exchange for food, money, or medicine. These networks facilitate the poaching or harvesting by providing the equipment necessary to commit such crimes. The indirect involvement of criminal networks, which do not engage in the poaching itself for instance, allows them to remain detached from the crime and evade authorities. Thus, at the local level, environmental crime generally involves the participation of low-income populations, which are exploited by criminal networks, suggesting a connection between poverty and environmental crime. At the national level, weak institutions expose countries to heightened risks of environmental crime. This is particularly true in fragile and developing states, where environmental crime is sometimes given insufficient priority, which provides opportunities for criminals to manipulate gaps. In addition to a weak legal framework to address environmental crimes, many authorities lack the expertise and the resources necessary to detect and prevent them. This may be further compounded by a lack of transparency within these authorities. Overall, poor enforcement capacity, coupled with inadequate legal systems and shaky institutions, provide an ideal setting for criminals to engage in environmental crime. At the global level, environmental crime "is driven and sustained by consumers who are willing to pay high prices for the commodity, regardless of its origin or legality" (UNEP, CITES, IUCN, & TRAFFIC 2013, p. 40). High demand is due either to the symbolic value or to the rare nature of some environmental commodities, which command exorbitant prices. Because they risk very little, yet stand to gain so much, criminals are increasingly drawn to environmental crime and to the traffic in environmental commodities (Interpol/UNEP Environment, 2016, p. 53).

Importantly, specific commodities (such as tiger skins, rhino horns, and elephant ivory) require an integrated network that links the point of origin to the point of sale, even though the gatherers/producers and consumers/clients may be disparate and unconnected. This demands local knowledge of poaching practices and communal contexts, as well as transit routes and venues for opportunistic exchanges (Wong, 2019). Particularly in relation to cross-border crime the operational success of criminal networks involves conscious, intentional criminality and where required, the paying of bribes to appropriate government officials in regards illegal logging, illegal wildlife and other environmental crime commodities. The corrupt engagement of government officials, including those at the front end of environmental law enforcement (e.g., customs officials, local police), is likewise part of the overall picture of environmental crime (INTERPOL/UNEP Environment, 2016).

Indeed, corruption and natural resource extraction have long gone hand-in-hand, especially in the wildlife and forestry sectors, and increasingly in relation to water supply. In these sectors corruption is evident in activities such as payment of bribes to government officials or politicians for preferential treatment, extortion by officials from operators to artificially legalise illegal operations, evading of national regulations with relative impunity, and bribing customs and border security personnel to ignore smuggling (Van Dinh, 2012; INTERPOL/UNEP Environment, 2016). The nature and extent of corruption will vary from country to country and situation to situation, as will the officials working in enforcement, detection, prosecution, the judiciary and policy making who are implicated in corruption (UNODC, 2012). Arguments to the contrary notwithstanding ('it does not happen here'), corruption is ubiquitous and global—that is, it is a Western as well as non-Western phenomenon (White, 2015). For example, moral and legal corruption is evident in places such as Australia, and especially in regards the natural resources extraction industries (White, 2016), including water management (see below). The specific dynamics of corruption may vary, but the core problem of corrupt business dealings is extensive.

Examples of different types of organized environmental crimes include the trafficking of animal and plant species; illegal poaching of flora and fauna; trafficking of ozone-depleting substances, hazardous chemicals, and harmful pesticides; waste crimes; fisheries crime; forestry crimes; bio-piracy and transportation of controlled biologically or genetically modified materials; illegal mining; and fuel trafficking, to avoid paying taxes (White, 2008; Environment Investigation Agency (EIA), 2008; Elliott, 2009; Eman, 2010; Interpol/UNEP Environment, 2016). Elliott (2009) has observed that environmental crime greatly expanded and become transnational when organized criminal groups took advantage of the benefits of economic liberalization and globalization of the world economy. The consequence has been an increase in the frequency and volume of cargo and shipments, reduced border controls, and the facilitation of the transfer of resources through the global financial and banking systems, which offer more options for money laundering and diversion of profits into legitimate business and company. The convergence of environmental crime with other crimes usually includes corruption, counterfeiting, drug trafficking, cybercrime, illegal immigration, financial crime, arms trafficking, terrorism, and human trafficking (Interpol/UNEP Environment, 2016). An example of this is human trafficking and persons being forced into virtual slavery on illegal fishing boats (UNODC, 2011). Furthermore, organized crime networks are involved in trading and transporting of more than one illegal commodity at the time. South and Wyatt (2011) note that such multi-commodity smuggling can involve the same trade routes and traffickers.

3 Water Crimes from an Organized Crime Perspective

Organized environmental crime often operates on the border of the lawful. The embeddedness of organized crime in legal sectors of the economy (van Uhm, 2016, p. 254) presents as a relevant contemporary issue when talking about crimes involving water. The line between legal and criminal is usually blurred, especially when legal and illegal actors interact with each other (e.g. Italia Mafia and corruption of officials). Van Uhm (2016, p. 255) notes that "white-collar crime and corporate crime may have connections with organized crime due to encounters between legal and illegal economies", and we believe that water sector at the very moment presents one of such potentially high-risk field.

Numerous violations by legitimate companies in the field of water privatisation—for example, Johannesburg Water in South Africa, Water and Sewage Corporation in Ghana, City Water in Tanzania, Coca Cola in Rajasthan, Aguas del Tunari in Bolivia, TAIPED in Athens, Veolia and Suez in Paris—also raises questions about collusion between corporations and organised criminal networks (Humar & Eman, 2017). Like in the water sector in Calabria (Sergi & South, 2016) and Napoli (Abadinsky, 2016), water is presenting as a new enormously profitable natural resource, especially due to the human need of it for survival.

Despite the fact that crimes against water are hard to detect, assess and prosecute, we can describe water crimes as having three aspects: an environmental resource damaged by a crime (e.g. surface water pollution, or fraudulent water quality reporting); the object of a crime (e.g. water theft or corruption on the part of private companies related to the economic control/exploitation of water); and the means of the crime (e.g. intentional flooding or the deliberate poisoning of a water supply). There may also be criminal threats to water management infrastructure (e.g. terrorism or cyber threats) (Water Crimes, 2016). Specific water crimes can include pilfering of water from pipelines, illegal waste management, water theft, river and marine pollution, manipulation of sampling methods to avoid treatment costs, fraud and illegal trafficking of water, terrorism and cyber threats to water management operations, illegal waste discharges from factories, unauthorised consumption from the water network, and water smuggling. (Kelly & Oldring, 2015; Moretti et al., 2017; Water Crimes, 2016).

Interpol (Interpol/UNEP Environment, 2016) highlights three groups of water crimes: (1) water fraud (e.g. the alteration of sampling techniques or results to avoid treatment costs that have negative health implications) (2) water pollution (i.e. intentional contamination of water), and (3) water theft (i.e. the unauthorized use and consumption of water before it reaches the intended end-user). It is estimated that between 30 and 50% of the global water supply is illegally purchased. Regions experiencing chronic water stress (e.g. Southern Europe, north east Africa) and marginalized deprived areas (e.g. slums in India, Bangladesh, and Brazil) are particularly vulnerable. Local communities are often forced to find alternative solutions to fulfil their daily needs for water. In Africa, the number of unregulated wells has skyrocketed from 2 million to an estimated 23–25 million in a decade. While this reflects poor water management, this practice is also conducive to the major degradation of water resources (Interpol/UNEP Environment, 2016, p. 33).

Organised criminal groups involved in water fraud and water theft jeopardize the integrity of the existing supply chain. What is more, water crimes intersect with other criminal activities, such as fraud, document forgery, corruption, bribery, and misappropriation of funds, and these also are associated with organized criminal groups.

Corruption in the water sector is also fuelled by the involvement of organized crime, and more specifically mafia-type groups. The monopoly over the water supply underpins the power and influence that organized crime groups have over vulnerable communities living in poor and/or marginalized areas (e.g., slums), particularly in Bangladesh, Brazil, and India (Interpol/UNEP Environment, 2016, p. 34).

The populations living in these places can access water only through a criminal group. This way, people from the slums are subject to the volatile nature of prices set by organized crime syndicates on resources which are essential for their survival and development. Interpol (Interpol/UNEP Environment, 2016, p. 58) emphasizes that the dependence on environmental crime, which is perceived as a survival option, not only turns poor communities into accomplices to the crime but also draws them into a life influenced by crime.

The uneven distribution of water around the world has led to a situation in which water theft is on the rise. The status of water as a tradeable commodity on the market has also increased the prices of water and placed it firmly in the gaze of organized crime. Water theft includes the illegal acquisition of natural watercourses as well as piped or harnessed water, both of which are designated as "non-revenue water", meaning water that is "lost" before reaching its intended consumer. Loss of water through theft can occur in a variety of ways including damaging or removing water meters and physically installing a connection to water distribution pipes as well as stealing aid tanks (The Global Initiative, 2014). Water scarcity and smuggling are often seen in North Africa (e.g. Nigeria, Mozambique, Kenya, Sudan, Chad) and the Middle East (e.g. Israel, Jordan) (The Global initiative, 2014), and from Canada to the United States (Kelly & Oldring, 2015).

Wyatt et al. (2018) observe that corruption associated with wildlife trafficking occurs within the context of three structural elements: where agents within the criminal justice system collude with criminal groups (for example, taking bribes to acquit smugglers or to permit them to cross borders); where there are weak financial institutions and failure to enforce money laundering and tax evasion laws; and where there is lack of political will to enforce laws and strong measures directed against wildlife trafficking. Similar observations can be seen in relation to water crimes, as recently demonstrated in Australia.

In July 2017, an Australian Broadcasting Corporation Four Corners investigation revealed a series of improper conducts pertaining to the Murray-Darling Basin, the largest fresh-water system in Australia (ABC, 2017). Four Corners is a long-running investigative current affairs television program produced by the national public broadcaster (the ABC). Specifically, it was alleged that in New South Wales:

- huge amounts of water were being diverted (stolen) for use by large agriculture companies upstream—more than they were entitled to (1.1 billion gigalitres for one property alone);
- there was pump tampering and failure to keep diaries and logs;
- the top regulator in New South Wales, Gavin Hanlon, had offered to help lobbyists campaign against the Murray-Darling Basin plan (by sending de-identified government documents to industry people); and
- the Strategic Investigations Unit in New South Wales was disbanded at precisely the time when it was asking for more resources to address significant problems (compliance was moved out of the relevant government department).

The broadcast led to no less than eight official inquiries into these issues (Murray-Darling Basin Authority, 2017; Craik, 2018), involving federal and state agencies, anti-corruption and water management groups, the New South Wales Ombudsman, and a South Australian Murray-Darling Basin Royal Commission.

At a concrete level, the actions of governments and businesses may privilege the rights and interests of the powerful over the public interest. Sometimes this takes the form of direct state-corporate collusion (state-corporate crime); in other instances, it involves manoeuvring by government officials or company executives to evade the normal operating rules of planning, development and environmental impact assessment. Certainly, the Four Corners program suggested this when it accused the top regulator, Gavin Hanlon, of helping lobbyists who were campaigning against the Murray-Darling Basin plan (Matthews, 2018).

Despite the huge interest in the water, as this case study illustrates, transparency in this sector is often limited, which opens the doors for corruption. The World Bank (2016) estimates that criminal engagement and/or corruption in the water sector causes between 20 and 40% of lost finances. Petty bribes to government officials and water suppliers are used to falsify meter readings, avoid disconnections, and conceal illegal connections. The rising demand for water only serves to reinforce these corrupt practices and enforce a sense of discretionary power among officials, consequently weakening a country's rule of law (The Global Initiative, 2014). Organized crime groups are also very resourceful in other ways when it comes to making a profit. This has happened in Calabria, Italy, where organised criminal networks have systematically misappropriated EU funds under the guise of enhancing the environment. For instance, in the years 2013 and 2014, the EU invested more than 133 million Euros in 22 environmental projects, of which 16 projects were in the water sector, one related to air and one to fire, and four projects related to the soil sector. Analysis of past projects revealed negative consequences that included water pollution (Sergi & South, 2016, p. 90). The water pollution included negligent or missing depuration processes, unauthorized dumping at sea, sewers directly connected to the sea, and oil spilling. In some locales, there is thus a systematic interpenetration of organized crime interests and malpractices, and regulatory omissions at the political and administrative levels (Legambiente, 2014; Ruggiero & South, 2010; Sergi, 2015).

4 Disrupting Organised Environmental Crime

Environmental crimes such as water theft are currently one of the most profitable forms of criminal activity (Walters, 2013, p. 281). They must be tackled at different levels using a variety of methods. For example, multilateral cooperation is important for the prevention of water smuggling (The Global Initiative, 2014). Yet, past experience show that environmental protection regimes and legislation often do not work due to implementation issues or practical barriers (Nurse, 2012; Wellsmith, 2011; White, 2012). Key issues in policing environmental crime include difficulties in detection, lack of specialist knowledge, and the involvement of a range of criminal actors (White & Heckenberg, 2014, pp. 222–223; Nurse, 2015, p. 126).

Environmental crime tends to be enforced reactively rather than proactively. One possible alternative to this is adoption of harm-oriented policing (Ratcliffe, 2015). The goal of harm-focused policing is to inform policing priorities by weighing up the harms of criminality, as well as crime and disorder. This would then be used in order to focus police resources in furtherance of both reducing crime and harm (Ratcliffe, 2015, p. 3). Given that police more often deal with environmental harm than environmental crime, such an approach would ideally help to change the standard police approach to environmental issues.

On the other hand, Nurse (2015) suggests the establishment of dedicated and professional environmental policing agencies for each specific field, such as fish and wildlife, logging, waste issues, and so on. As policing is often faced with lack of expert knowledge, multi-agency cooperation would be very welcome at this point. Indeed, the multi-dimensional nature of environmental crimes offers the possibility for the use of innovative means to combat organised criminal groups. Dighe and Pettus (2011), for example, describe the use of 'environmental enforcement sweeps' that involve the use of administrative, civil, and criminal enforcement tools all at the same time. In a similar vein, Barret and White (Barrett & White, 2017) talk about multi-agency collaboration and describe it as a key means of responding to organized crime groups and networks. In each case the idea is, that from a law enforcement perspective, it is more effective to focus on 'disrupting crime'. By focusing on disruption, law enforcement agencies are able to use combined tactics that impede the ability of organized crime groups to function. This is based on the recognition that environmental offenses, especially organized environmental crime, typically include other criminal offenses or illegal activities, such as bribery, threats, extortion, corruption, and falsification of documents.

Collaborations will differ according to each crime's characteristics. For example, in a Durham case involving a local organised criminal group, multi-agency cooperation involved the EPA, the municipality, insurance companies, police, health and safety agency, fire safety agency, general register, insolvency practitioners' agency, UK passport office, electricity company, water supply company, driver vehicle standards agency, and politicians (Barrett & White, 2017). The specialty of the Durham case was that first the activities of the organized crime group were assessed with the goal to identify all possible wrongdoings. Secondly, relevant experts from various agencies were brought together in a working group to address identified anomalies. Thirdly, the plan of the coordinated attack on the organized crime group included the participation of the key agencies. The agencies acted in a coordinated manner with the common goal to undermine the ability of organized crime group to function (Barrett & White, 2017).

Despite the use of new intelligence-led detection and investigation policing methods, police still have to target the high-level players in organized criminal networks in order to break them up and achieve wide-ranging impact (Akella & Allan, 2014). Due to the fact that responding to organized criminal networks has to include robust action across the various dimensions of the enforcement chain, law enforcement agencies and relevant stakeholders on all levels, local, regional, national and global, must be included. The prevention of environmental crime is enhanced through the cooperation of and connection between national and international law enforcement agencies, NGOs and governments (Faure, De Smedt, & Stas, 2015; Pink & White, 2016).

For example, the National Environmental Security Taskforce (NEST) intervention model is one of the possible responses to organized environmental crime. Initiated by Interpol, the NEST refers to multi-agency cooperation in the environmental sector. It operates as a task force comprised of a team of experts who work together to address specific issues and enables "an integrated multi-pronged approach to environmental law enforcement, working with many partner agencies and NGOs which share an interest in fighting environmental crime" (White, 2016, p. 95). Good examples of the NEST's use are Interpol's projects Leaf, Scale, Wisdom, Predator, and Eden that deal with crimes such as illegal fishing, pollution offences and wildlife trafficking (Interpol/UNEP Environment, 2016). Successfully responding to organized environmental crime and their networks requires coordinated efforts at the local, regional, national and global level, together with strong inter- and intra-agency collaboration. For this, there is need for political will to provide the needed resources, as well as combat systemic corruption within the natural resource sectors.

5 Conclusion

The main reasons for the involvement of organized criminal groups in water crime are high profits, low penalties and lack of attention from law enforcement agencies. The water sector is also marred by corruption and systemic failures of regulation. Yet, environmental crime is a growing and devastating threat that destabilizes the entire Planet's ecosystem. It is clear that more needs to be done to address not only water crimes as such, but the role of organised criminal groups in perverting the supply and quality of water.

Interpol suggests that solutions will require a number of interrelated elements: a comprehensive and multidisciplinary approach in tackling organized environmental crime and its convergence with other criminal activities; an increase of multi-agency information exchange; an increase of global focus on the implementation of environmental protection policy; and stronger financial support for regulators and environmental law enforcement (Interpol/UNEP Environment, 2016). Innovative responses to the crimes associated with water crimes are certainly needed. As part of this, it is essential to disrupt organized crime groups in multiple ways, using a wide range of administrative, civil and criminal measures. Importantly, this disruption needs to include community participation and grassroots initiatives, since frequently it is the most vulnerable who suffer the most from the nefarious activities of organised crime groups. Water is a human right. To guarantee this, it must be protected and defended as such.

References

Abadinsky, H. (2016). *Organised crime* (9th ed.). Boston: Cengage Learning.

Akella, A., & Allan, C. (2014). *Dismantling wildlife crime: Executive summary*. Washington: World Wildlife Fund.

Alston, M., & Mason, R. (2008). Who determines access to Australia's water? Social flow, gender, citizenship and stakeholder priorities in the Australian water crisis. *Rural Society, 18*(3), 214–219.

Australian Broadcasting Corporation (ABC) (2017). Digging into Adani. Four Corners, 2 October. Retrieved 3 December, 2019, from https://www.abc.net.au/4corners/digging-into-adani/9008500

Ayling, J. (2013). What sustains wildlife crime? Rhino horn trading and the resilience of criminal networks. *Journal of International Wildlife Law Policy, 16*(1), 57–80.

Barrett, S., & White, R. (2017). Disrupting environmental crime and the local level: An operational perspective. *Palgrave Communications, 3*, 2. Retrieved 27 March, 2017, from https://www.nature.com/articles/s41599-017-0006-3

Bisschop, L. (2011). Transnational environmental crime: Exploring (un)charted territory. In M. Cools, B. de Ruyver, M. Easteon, L. Pauwels, P. Ponsaes, T. V. Beken, et al. (Eds.), *EU criminal justice, financial & economic crime: New perspectives, governance of security research papers* (Vol. 5, pp. 155–183). Antwerpen: Maklu.

Craik, W. (2018). *Report of the independent reviewer: Basin compliance compact.* Retrieved 13 January, 2020, from https://www.mdba.gov.au/sites/default/files/pubs/report-independent-reviewer-mdba-compliance-compact_1.pdf

Dighe, K., & Pettus, L. (2011). Environmental justice in the context of environmental crime. *United States Attorneys' Bulletin, 59*(4), 3–14.
Dobovšek, B. (1997). *Organizirani kriminal*. Ljubljana: Unigraf.
Dobovšek, B. (2009). *Transnacionalna kriminaliteta*. Ljubljana: Fakulteta za varnostne vede.
Elliott, L. (2007). Transnational environmental crime in the Asia Pacific: an 'un(der)securitized' security problem? *The Pacific Review, 20*(4), 499–522.
Elliott, L. (2009). Combating transnational environmental crime: Joined up thinking about transnational networks. In K. Kangaspunta & I. H. Marshall (Eds.), *Eco-crime and justice: Essays on environmental crime* (pp. 55–77). Torino: UNICRI.
Eman, K. (2010). Organizirana ekološka kriminaliteta—nekoč teoretični pojem, danes preteče dejstvo. In *Zbornik 3. konference kazenskega prava in kriminologije* (pp. 95–104). Ljubljana: GV založba.
Eman, K., Furdi, T., Hacin, R., & Dobovšek, B. (2016). A discourse on the gray economy, corruption, and organized crime in Slovenia. In M. Edelbacher, P. C. Kratcoski, & B. Dobovšek (Eds.), *Corruption, fraud, organized crime, and the shadow economy (Advances in police theory and practice series)* (pp. 167–180). Boca Raton/London/New York: CRC Press.
Eman, K., Kuhar, S., & Meško, G. (2017). Water crimes and policing. *Varstvoslovje, 19*(2), 105–119.
Environment Investigation Agency (EIA). (2008). *Environmental crime: A threat to our future*. Retrieved 20 March, 2017, from http://www.eia-international.org/files/reports171-1.pdf
Faure, M., De Smedt, P., & Stas, A. (Eds.). (2015). *Environmental enforcement networks: concepts, implementation and effectiveness*. Cheltenham: Edward Elga.
Federal Bureau of Investigation (FBI). (2010). *Glossary of terms*. Retrieved 27 March, 2017, from http://www.fbi.gov/about-us/investigate/organizedcrime/glossary
Gleick, P. (1998). Water in crisis: Paths to sustainable water use. *Freshwater Systems, 8*(3), 571–579.
Hayman, G. & Brack, D. (2002). *International environmental crime: The nature and control of environmental black markets*. Retrieved 29 November, 2011, from http://ec.europa.eu/environment/docum/pdf/02544_environmental_crime_workshop.pdf
Humar, S., & Eman, K. (2017). Privatizacija vodnih virov—nova oblika kriminalitete zoper vode. *Revija za kriminalistiko in kriminologijo, 68*(2), 116–134.
Interpol/UNEP Environment. (2016). *Strategic report: Environment, peace and security—A convergence of threats*. Retrieved 21 March, 2017, from https://bigcatrescue.org/wp-content/uploads/2016/12/INTERPOL-UNEP-Strategic-Report-Environment-Peace-and-Security-A-Convergence-of-Threats.pdf
Johnson, H., South, N., & Walters, R. (2015). The commodification and exploitation of fresh water: Property, human rights and green criminology. *International Journal of Law, Crime and Justice, 44*, 146–162.
Kelly, P. & Oldring, P. (2015). *Meet a Canadian making thousands of dollars smuggling water into the US*. Retrieved 27 March, 2017, from http://www.cbc.ca/radio/thisisthat/school-real-estate-word-search-champion-water-smugglers-1.3194224/meet-a-canadian-making-thousands-of-dollars-smuggling-water-into-the-u-s-1.3194276
Klenovšek, A., & Meško, G. (2011). International waste trafficking—Preliminary explorations. In G. Meško, D. Dimitrijević, & C. B. Fields (Eds.), *Understanding and managing threats to the environment in South Eastern Europe* (pp. 79–100). Dodrecht: Springer.
Lin, J. (2005). Tackling Southeast Asia's Illegal Wildlife Trade. *Singapore Year Book of International Law, 9*(1), 191–208.
Legambiente. (2014). *Ecomafia 2014*. Napoli: Marotta e Cafiero Editori.
Massari, M., & Monzini, P. (2004). Dirty business in Italy: A case-study of illegal trafficking in hazardous waste. *Global Crime, 6*(3/4), 285–304.
Matthews, K. (2018). *Final Report: Independent investigation into NSW water management and compliance*. Retrieved 13 January, 2020, from https://www.industry.nsw.gov.au/__data/assets/pdf_file/0019/131905/Matthews-final-report-NSW-water-management-and-compliance.pdf

Moretti, F., Segato, L., Capello, N., Mattioli, W., Monnier, C., Aleksic, D., et al. (2017). *European report on water crimes*. Szentendre: The Regional Environmental Center for Central and Eastern Europe.

Murray-Darling Basin Authority. (2017). 2017 Basin Plan Evaluation. Accessed January 13, 2020, from https://www.mdba.gov.au/basin-plan-roll-out/monitoring-evaluation/2017-basin-plan-evaluation

Newman, J. (2001). The tricks of illegal trade: How criminals smuggle ODS. In M. Rynn (Ed.), *Illegal trade in ozone depleting substances: Is there a hole in the Montreal protocol* (pp. 14–19). Paris: UNEP.

Nurse, A. (2012). Repainting the thin green line: The enforcement of UK wildlife law. *Internet Journal of Criminology, 12*, 1–20.

Nurse, A. (2015). *Policing wildlife: Perspectives on the enforcement of wildlife legislation*. Hampshire: Palgrave Macmillan.

Pink, G., & White, R. (2016). Collaboration in combating environmental crime: Making it matter. In G. Pink & R. White R. (Eds.), *Environmental crime and collaborative state intervention* (pp. 3–19). Basingstoke: Palgrave Macmillan.

Ratcliffe, J. H. (2015). Harm-focused policing. *Ideas in American Policing, 19*, 1–11.

Ruggiero, V., & South, N. (2010). Green criminology and dirty-collar crime. *Critical Criminology, 18*(4), 251–262.

Schmidt, C. W. (2004). Environmental crimes: Profiting at the Earth's expense. *Environmental Health Perspectives, 112*(2), 96–103.

Sergi, A. (2015). Mafia and politics as concurrent governance actors. Revisiting political power and crime in Southern Italy. In P. C. van Duyne, A. Maljević, G. A. Antonopoulos, J. Harvey, & K. von Lampe (Eds.), *The relativity of wrongdoing: Corruption, organised crime, fraud and money laundering in perspective* (pp. 43–70). Oisterwijk: Wolf Legal Publishers.

Sergi, A., & South, N. (2016). Earth, water, air, and fire. environmental crimes, mafia power and political negligence in Calabria. In G. Antonopolous (Ed.), *Illegal entrepreneurship, organised crime and social control: Essays in honour of Prof. Dick Hobbs* (pp. 85–100). New York: Springer.

Siegel, L. J. (2001). White-collar and organized crime. In S. Horne (Ed.), *Criminology: Theories, patterns and typologies* (pp. 123–146). Wadsworth: Thomson Learning.

Situ, Y., & Emmons, D. (2000). *Environmental crime: The criminal justice system's role in protecting environment*. London, New Delhi: Sage Publications.

South, N., & Wyatt, T. (2011). Comparing illicit trades in wildlife and drugs: An exploratory study. *Deviant Behaviour, 32*, 538–561.

Spapens, T., White, R., van Uhm, D., & Huisman, W. (Eds.). (2018). *Green crimes and dirty money (green criminology)*. Abingdon: Routledge.

Stølsvik, G. (2010). *The transnational nature of illegal fish*ing. Retrieved 3 November, 2010, from http://www.financialtaskforce.org/2010/09/27/the-transnational-nature-of-illegal-fishing/

Tacconi, L. (2008). The Problem of Illegal Logging. In L. Tacconi (Ed.), *Illegal logging: Law enforcement, livelihoods and the timber trade* (pp. 1–16). London: Earthscan.

The Global Initiative. (2014). *Global Initiative against transnational organised crime*. Retrieved 27 March, 2017, from http://globalinitiative.net/water-smuggling/

UNODC. (2011). *Guidelines for the Safe handling and disposal of chemicals used in the illicit manufacture of drugs*. Retrieved 19 November, 2019, from https://www.unodc.org/documents/southeastasiaandpacific//Publications/2011/Disp.Manual_English.pdf

UNODC. (2012). *Amphetamine type stimulants in Vietnam: A review of the availability, use and implications for health and security in Vietnam*. Retrieved 19 November, 2019, from https://www.unodc.org/documents/southeastasiaandpacific//2012/05/ats-report-vietnam/ATS_EN_Final_10042012_-_reduced_size.pdf

UNESCO. (2012). *Report on the findings of phase one of the UNESCRO-WWAP: Water scenarious to 2050*. Retrieved 27 March, 2017, from http://www.unesco.org/new/en/natural-sciences/environment/water/wwap/world-water-scenarios/phase-1/

UNEP, CITES, IUCN, & TRAFFIC (2013). *Elephants in the Dust - The African Elephant Crisis. A Rapid Response Assessment.* Arendal: UNEP.

United Nations Convention Against Transnational Organised Crime (2000). Retrieved 27 March, 2017, from https://www.unodc.org/documents/middleeastandnorthafrica/organised-crime/united_nations_convention_against_transnational_organized_crime_and_the_protocols_thereto.pdf

Van Dinh, T. T. (2012). Addressing corruption in the environmental sector: How the United Nations Convention against Corruption provides a basis for action. In United Nations Office on Drugs and Crime (ed.), *Corruption, Environment and The United Nations Convention Against Corruption* (pp. 34–50). Vienna: United Nations Office on Drugs and Crime.

van Uhm, D. P. (2016). *Uncovering the illegal wildlife trade: Inside the world of poachers, smugglers and traders*. New York: Springer.

Walters, R. (2013). Eco Crime. In J. Muncie, D. Talbolt, & R. Walters (Eds.), *Crime: Local and global*. Cullompton: Willan Publishing.

Water Crimes. (2016). *Water crimes project.* Retrieved 21 March, 2017, from http://www.water-crimes.eu/

Watson, M. (2005). Organised crime and the environment: The British experience. *European Environmental Law Review, 14*(8/9), 207–213.

Wellsmith, M. (2011). Wildlife crime: The problems of enforcement. *European Journal on Criminal Policy and Research, 17*(2), 125–148.

White, R. (2008). *Crimes against nature: Environmental criminology and ecological justice*. Cullompton, Portland: Willan Publishing.

White, R. (2012). NGO engagement in environmental law enforcement: Critical reflections. *Australian Policing, 4*(2), 7–12.

White, R. (2015). Climate change, ecocide and the crimes of the powerful. In G. Barrak (Ed.), *The Routledge international handbook of the crimes of the powerful* (pp. 211–222). Abingdon: Routledge.

White, R. (2016). Building NESTs to combat environmental crime networks. *Trends in Organised Crime, 19*(1), 88–105.

White, R., & Heckenberg, D. (2014). *Green criminology: An introduction to the study of environmental harm*. London: Routledge.

Wong, R. W. Y. (2019). *The Illegal Wildlife Trade in China Understanding The Distribution Networks.* Hampshire: Palgrave Macmillan.

World Bank. (2016). *Water overview.* Retrieved 27 March, 2019, from http://www.worldbank.org/en/topic/water/overview

Wyatt, T., Johnson, K., Hunter, L., George, R., & Gunter, R. (2018). Corruption and Wildlife Trafficking: Three Case Studies Involving Asia. *Asian Journal of Criminology, 13*(1), 35–55.

Chapter 5
Building the Water Crimes Inventory

Walter Mattioli, Nicola Capello, and Lorenzo Segato

1 Introduction

Water is a fundamental natural resource for life. In time, it has become a "hot product" attractive for criminals (Clarke, 1999), potentially exposed to the illicit interests of individuals, companies, terrorists, and organised crime. Nevertheless, crimes against water remain difficult to assess, detect and prosecute, especially because legislation and policies on freshwater vary greatly from country to country, data are inconsistent or not comparable, and water crimes are often recorded under other offences - like fraud, corruption, theft, forgery/counterfeiting of documents, terrorism, or other crimes against health or the environment (Segato et al., 2017). In addition to this, academic and law enforcement interest in water crime issues (Bricknell, 2010; Brisman, McClanahan, & South, 2016; White, 2009) is very recent.

Water crimes suffer the same complexity and ambiguity of any other environmental crime. The term "environmental crime" is often used as an umbrella term for crimes related to several topics: biodiversity, wildlife, animals, natural resources, hazardous waste, banned substances, and environmental quality. Therefore, there is no clear, commonly accepted definition of water crimes, but only some general observations (Brisman & South, 2019; Segato et al., 2017). Water as an environmental resource can either be damaged by a crime (e.g. surface water pollution, or fraudulent water quality reporting), be the target of a crime (e.g. drinking water

W. Mattioli (✉)
CREA - Research Centre for Forestry and Wood, Arezzo, Italy

Department for Innovation in Biological, Agro-food and Forest systems (DIBAF), University of Tuscia, Viterbo, Italy
e-mail: walter.mattioli@crea.gov.it; walter.mattioli@unitus.it

N. Capello · L. Segato
Research and Action - REACT, Padova, Italy
e-mail: nicola.capello@re-act.it; lorenzo.segato@re-act.it

K. Eman et al. (eds.), *Water, Governance, and Crime Issues*,
https://doi.org/10.1007/978-3-030-44798-4_5

theft) or be the means of a crime (e.g. intentional flooding or the deliberate poisoning of a water supply). There may also be criminal threats to water management infrastructure such as terrorism and cyber-attacks (Segato et al., 2017).

Although crime against water was considered years ago as an emerging global issue (INTERPOL, 2012) and water management as an emerging environmental crime (INTERPOL, 2016), no consistent attempts have been made to categorise different types of water crimes or to compile a comprehensive water crimes inventory in order to fill the gap about the nature and extent of these kinds of activities, which are still relatively unknown.

Against this background, the Water Crimes Project (WCP, 2016) carried out wide-ranging and systematic research into crimes against water and provided the first strategic analysis of offences and threats related to the freshwater sector in Europe, for the purpose of creating the very first inventory of water-related crimes. The WCP has involved INTERPOL and UNODC in drafting a definition of the water crimes. This paper reports on the process, the methodology and the first results of this attempt at building a water crimes inventory.

2 The International Legal Framework

One of the main challenges in water crime analysis is the absence of a legal definition of these offences. In the following, we present a review of the existing international legal instruments concerning the water sector which bear potential relevance for the definition of water and water-related criminal offences:

- *UN Convention on the Law of Non–Navigational Uses of International Watercourses, 1997 (UN Watercourses Convention; entry into force: 2014).* This convention adopts a wide definition of the uses of international watercourses for purposes other than navigation, with the aim of protecting, preserving and managing these waters. An "international watercourse" is defined as a system of surface water and groundwater constituting a unitary whole and situated in different states. Although the convention seeks for commitments to prevent and reduce pollution in this type of water, there is no reference to the need to include water or water-related crimes in the legislation of each state;
- *Convention on the Protection and Use of Transboundary Watercourses and International Lakes, 1992, adopted by the United Nations Economic Commission for Europe (UNECE Water convention; entry into force: 1996).* This convention focuses on transboundary surface water and groundwater bodies that are located on the boundaries between two or more states. Its scope is to promote cooperation related to the management of such water bodies, highlighting the need to prevent pollution and to achieve sustainable development that protects ecosystems. Like the UN Watercourses Convention, it focuses on cooperation among different states on water management, with the declared aim of preventing pollution and protecting the environment. There are no provisions regarding the need to establish any criminal offences in this field;

- *Directive 2008/105/EC of the European Parliament and of the Council of Dec, 16, 2008, on environmental quality standards in the field of water policy, amending and subsequently repealing council directives 82/176/EEC, 83/513/EEC, 84/156/EEC, 84/491/EEC, 86/280/EEC and amending directive 2000/60/EC of the European Parliament and of the Council (EU Water Quality Directive).* This directive establishes the so-called environmental quality standards (EQS) for priority substances and other pollutants, with the objective of guaranteeing good surface water chemical status. It derives from the EU Water Framework Directive (see below: Directive 2000/60/EC of the European Parliament and of the Council establishing a framework for Community action in the field of water policy), and, though Directive 2008/105/EC, amends this last directive, adopting its concepts and definitions. It provides a list of different substances, determining the maximum permissible amount for different kinds of surface waters in order to guarantee the safety and security of water resources. The purpose of the directive is thus to avoid the chemical pollution of water resources, considering the dangerous effects also on human health;
- *Directive 2008/56/EC of the European Parliament and of the Council of June 17, 2008, establishing a framework for community action in the field of marine environmental policy (EU Marine Strategy Framework Directive).* This directive establishes objectives for the protection and conservation of the marine environment against harmful human activities; it contains different strategies to protect marine ecosystems and to ensure that economic activities linked to the marine environment are sustainable;
- *Council Directive 91/271/EEC concerning urban wastewater treatment.* The aim of this directive is to protect the environment from the adverse effects of urban wastewater discharges and discharges from certain industrial sectors. The directive seeks to establish the regulation and monitoring of the collection and treatment of wastewater and to prevent water pollution;
- *Council Directive 91/676/EEC concerning the protection of waters against pollution caused by nitrates from agricultural sources.* The goal of this directive is to protect water quality by preventing nitrates from agricultural sources polluting groundwater and surface waters, and by promoting the use of good farming practices. States should identify waters that have been polluted or that are at risk of pollution caused by nitrates and should ensure that good agricultural practices can be implemented to prevent and control this kind of pollution;
- *Council Directive 98/83/EC on the quality of water intended for human consumption* (*Drinking Water Directive*). The goal of this directive is to protect human health from the adverse effects of the contamination of water intended for human consumption. All member States must implement certain measures to guarantee that water for human consumption is wholesome and clean. Regular monitoring of this kind of water is crucial in order to guarantee its safety and protect human health;
- *Directive 2000/60/EC of the European Parliament and of the Council establishing a framework for community action in the field of water policy* (*EU Water Framework Directive*): this is the basic legal instrument for implementing and

developing a European water policy. As a framework directive, its scope is to establish a framework for the protection of inland surface waters, transitional waters, coastal waters and groundwater, and to promote the implementation of measures and strategies to protect all these types of water and to prevent and control pollution;

- *Directive 2006/7/EC of the European Parliament and of the Council of February 15, 2006, concerning the management of bathing water quality and repealing Directive 76/160/EEC* (*Bathing Water Directive*). This directive refers to surface waters that can be used for bathing (except for swimming pools and spa pools), confined waters subject to treatment or used for therapeutic purposes, and confined waters artificially separated from surface waters and groundwater. Its aim is to guarantee water quality by establishing related management measures and parameters;
- *Directive 2006/118/EC of the European Parliament and of the Council on the protection of groundwater against pollution and deterioration*. This directive sets groundwater quality standards and introduces measures to prevent or limit the pollution of this kind of water. The directive describes measures to prevent or limit the introduction of pollutants into groundwater;
- *Directive 2008/99/EC on the protection of the environment through criminal law*. The aim of this directive is to guarantee that the most serious offences against the environment are punished by criminal sanctions. It obliges member states to define as criminal offences different incidents, whether intentional or at least occurring due to serious negligence, that result in relevant damage or a threat to different elements of the environment. Water is considered as one of the elements that can suffer damage due to the different kinds of behaviour mentioned in the directive. Nevertheless, the reference to criminal sanctions is also very general, as the directive calls for "effective, proportionate and dissuasive criminal penalties" but establishes neither the nature nor the minimum duration of the sanction;
- *Council Directive 2008/114/EC on the identification and designation of European critical infrastructures and the assessment of the need to improve their protection*. Albeit not specifically related to water or the environment, this directive is relevant because it aims to guarantee the protection of "critical infrastructures", defined as "an asset, system or part thereof located in member states, which is essential for the maintenance of vital societal functions, health, safety, security, economic or social well-being of people, and the disruption or destruction of which would have a significant impact in a member state as a result of the failure to maintain those functions." Water infrastructure is essential for the correct distribution and availability of water for the population; thus, its disruption or destruction would create a great risk or damage to water accessibility. In this way, any mechanism for the preventive protection of such infrastructure would be an effective way to prevent these specific categories of water crimes;
- *Directive 2010/75/EU of the European Parliament and of the Council on industrial emissions* (*integrated pollution prevention and control*). Although not specifically focussed on water, this directive is important as it develops a general and

comprehensive framework for controlling industrial emissions and for preventing and controlling pollution coming from different sources. It replaces the previous Directive 2008/1/EC of the European Parliament and of the Council concerning integrated pollution prevention and control. The directive takes as a basis an "integrated permit". This means that the permits issued to industries take into consideration each agent's complete environmental performance, in order to avoid pollution being shifted from one medium (air, water or land) to another.

3 About Water Crimes

Directive 2008/99/EC on the protection of the environment through criminal law acknowledges the rise in environmental offences, and "obliges the Member States to provide for criminal penalties in their national legislation in respect of serious infringements of [a list of] provisions of Community law on the protection of the environment". Member states must criminalise at least nine examples of conduct that have an impact on the environment - namely, that cause (or are likely to cause) "death or serious injury to any person or substantial damage to the quality of air, the quality of soil or the quality of water, or to animals or plants", and only "when unlawful and committed intentionally or with at least serious negligence" (Article 3).

Although there is no universally agreed definition, "environmental crime" is generally understood to describe illegal activities harming the environment and aimed at benefiting individuals or groups or companies through the exploitation of, damage to, trade or theft of natural resources, including serious crimes and transnational organized crime (UNEP/INTERPOL, 2016).

For a long time, five areas have been considered of major importance in relation to environmental crimes: illegal trade in wildlife; illegal logging and its associated timber trade; illegal, unreported and unregulated fishing; illegal trade in controlled chemicals (including ozone-depleting substances) and illegal disposal of hazardous waste (INTERPOL, 2012); references to "water" were often missing. For instance, green criminologists have long referred to water pollution as a general example of a green crime (Eman, Meško, Dobovšek, & Sotlar, 2013; Gibbs & Boratto, 2017). The main issue is the absence of a common definition of water crimes, water-related crimes and water security. In fact, water crimes can range from pilfering of water from pipelines to illegal waste management by a factory, up to pollution or terrorist threats (Segato et al., 2017). Green criminologists have devoted little attention to the study of water pollution and its consequences (Johnson, South, & Walters, 2015; Lynch & Stretesky, 2013; McClanahan, 2014), leaving this area of research open for further analysis concerning the scope, consequences, regulation and punishment of water pollution crimes (Lynch, Barrett, Stretesky, & Long, 2017). In particular, much is left to be explored in terms of the scope of water pollution problems in different contexts; the analysis of water pollution violations and the enforcement of water pollution/quality regulations; the consequences of water pollution on ecosystem stability and waterway species; and the intersection of water pollution and environmental justice (Lynch, Long, Stretesky, & Barrett, 2017).

Recently, ESA (2019) in cooperation with INTERPOL has identified three main types of water-related crimes: (1) water fraud, which involves the alteration of sampling techniques or results to avoid treatment costs; the main danger connected with these practices are the negative health implications; (2) water pollution, which refers to the intentional contamination of water, usually by companies or vessels (oil-pollution crimes include illegal oil discharges, false statements or records, and bypassing pollution prevention equipment); (3) water theft, that is, the unauthorized use and consumption of water before it reaches the intended end-user. This classification supposedly derives from the work carried out during the WCP (2016).

Practically speaking, water crime does not enjoy a specific status among environmental crimes, mainly because of the following issues: the absence of an international classification of water-related crimes; no working definition of water-related crimes; the scarcity of criminological studies; no maps of criminal behaviour, motivation or threats; the complexity of data collection; and the lack of a systematic analytical approach. Consequently, water-related crimes are often recorded under other offences - like fraud, corruption, trafficking, falsification of documents, terrorism, etc.

For the International Classification of Crime for Statistical Purposes (ICCS) (UNODC, 2016) water crimes and water-related crimes might fall into the categories of acts against the natural environment, acts leading to death or intending to cause death, acts leading to harm or intending to cause harm to the person, involving fraud, deception or corruption, acts against public safety and state security, acts against property only.

4 Materials and Methods

When elaborating a comprehensive definition of water crimes, it should be considered that water may be: the environmental resource damaged by a crime (e.g. surface water pollution, or fraudulent water quality reporting); the object of a crime (e.g. water theft or corruption on the part of private companies related to the economic control/exploitation of water); or the means of committing a crime (e.g. intentional flooding or the deliberate poisoning of a water supply). Gleick (2006) emphasises that the biggest problem is that water resources and systems are attractive targets for criminals because there is no substitute for water, which creates opportunities for several types of illegal activities. The first step in building a water crimes inventory is the adoption of a universal working definition of water crimes including all meaningful forms of offences against water bodies, based on - but not necessarily limited to - the provisions contained in the criminal law of different countries. For this purpose, the following working definitions defined by Segato et al. (2017) have been used: "water crime" is defined as any punishable contravention or violation of the limits on human behaviour as imposed by national criminal legislation, against surface, and groundwater, or against water services, which also includes any intentional act that can potentially harm or damage water; "water-related crime", on the other hand, is defined as any punishable contravention or violation of the limits on human behaviour as imposed by national criminal legislation, which uses surface water and groundwater, or water services, as a means of committing other crimes.

WCP (2016) collected data on 86 water and water-related crimes in Hungary, Italy, Slovenia, Spain, and other EU countries in order to create a water crimes inventory.

Data from other EU countries were collected through a survey of the websites of national institutions and of publicly available official statistics.

Results show that the most common water crime cases identified were water pollution and water theft (losses in the public water supply system). These cases amount to 86% across EU countries. Surface water was the target in 36% of cases, water services in 33%, and groundwater in 12%. In 47% of cases, the water was intended for human consumption, in 7% for industrial use, and in 6% for agricultural purposes.

Water is mostly at risk before it enters the supply chain (38% of all cases) and in the distribution phase (28%). In terms of location, 43% of offences occurred in an urban context while 42% in a natural or rural context. There were no cyber offences recorded.

Offences against the environment are prevalent (76% of cases; of which 70% were against water quality, 26% against water services, and 4% against water quantity). Only 7% of cases were fraud, and 4% were offences against property.

In most cases (30%) the perpetrator was an enterprise; in 28% of cases the perpetrator was unknown; in 23% of cases the offender was an individual; in 11% of cases the perpetrator was the state or an institution; in 4% of cases it was another legal person; in 3% of cases an organised crime group; and in 1% of cases a criminal group.

In 70% of cases the offence was committed intentionally, and in 30% it was due to negligence. In 32% of all cases, the damage was in the form of water pollution; in 23% it was water abuse; in 21% it was related to water accessibility; in 6% it was related to water availability; in 16% it was water fraud; and in 5% it was some other kind of damage.

The collected case studies were then classified based on the existing crime classification systems. Five approaches were considered:

1. *legal approach*: this approach is based on a legal definition of offences, be they penal, civil, or administrative (strict definition of crime). Within this approach, data could be collected and classified according to different legal provisions in relevant member states; however, the applicability of this approach is severely hindered by the absence of a normative framework concerning water crimes. The classification of environmental crimes (e.g., from Directive 2008/PP/EC) includes only a few specific activities (art. 3) that impact on water, and no distinction is made between salted and fresh water;

2. *offender approach*: this approach is based on "perpetrator or offender", "*modus operandi*" or "motivation". Within this approach, data could be collected and classified according to the nature and motives of crimes. The analysis of the collected data has shown that such information is not sufficient to discriminate different categories of crimes against water, since many offences have an economic motivation, and very often the offenders remain unknown.
3. *target approach*: this approach is based on the type of water being the target of the crime, the means of the crime or its indirect victim. This approach may underpin preventive actions, as it helps to identify the water stages in which risks are higher. However, due to extension of the potential cascading effects, and the complexity of establishing influencing factors and impacts, this approach appears too complex for crime prevention, and it does not allow a simple and "friendly" classification of water crimes;
4. *process approach*: this approach is based on the water supply chain, as defined in the Directive 2000/60/EC, or water status. The analysis of case studies highlights the difficulty of finding a single phase of a water supply chain involved, because the water supply chain should be regarded not as a set of distinct phases, but rather as a continuous cycle;
5. *statistical approach*: this traditional approach is based on existing data collection systems for crimes. It aims at providing information that can be easily understood and allowing for exhaustive and unambiguous attributions (as shown in Table 5.1).

5 Water Crimes Classification

A tentative classification was attempted by applying the standard crime classification system - the UNODC ICCS (2016). The ICCS is a standard international tool to classify any acts that are unlawful and committed intentionally, or with at least serious negligence, that can harm people, objects, institutions or the environment. The ICCS is structured according to 11 main categories (Level 1), each including a number of subcategories (Fig. 5.1).

The case studies collected could be classified according to the following ICCS Level 1 categories (in order of relevance) as: 10 - Acts against the natural environment; 1 - Acts leading to death or intending to cause death; 2 - Acts leading to harm or intending to cause harm to the person; 7 - Acts involving fraud, deception or corruption; 9 - Acts against public safety and state security; and 5 - Acts against property only.

According to this tentative classification, several codes are apparently applicable to different cases: in 25% of cases, the same crime was classified in two or three different Level 1 categories. Only 34% were classified with a unique code; 39% were classified with two codes; 17% with three codes; 7% with four codes; and 2.5% with five codes.

Table 5.1 A tentative water crimes inventory based on the statistical approach

Water crimes inventory with statistical approach		Motivation				
		Economic	Terrorism	Political	Need	Conflict
Perpetrator	State	State corruption		Abuse of police force Torture		Well/river poisoning Infrastructure destruction Downstream drying
	Enterprise	Illegal water extraction Pollution Corruption Fraud				
	Individual	Fraud Pilfering Cybercrimes Corruption			Water theft Violence	
	Criminal group	Extortion	Cybercrimes Chemical, biological, radiological, nuclear, and explosive attacks Water infrastructure destruction Intentional flooding	Assaults Sabotage Propaganda		Intentional flooding Poisoning
	Organised crime	Extortion Corruption Violence				

	LEVEL 1 CATEGORIES
1	Acts leading to death or intending to cause death
2	Acts leading to harm or intending to cause harm to the person
3	Injurious acts of a sexual nature
4	Acts against property involving violence or threat against a person
5	Acts against property only
6	Acts involving controlled psychoactive substances or other drugs
7	Acts involving fraud, deception or corruption
8	Acts against public order, authority and provisions of the State
9	Acts against public safety and state security
10	Acts against the natural environment
11	Other criminal acts not elsewhere classified

Fig. 5.1 UNODC ICCS categories: Level 1 (UNODC ICCS, 2016)

Water crimes and water-related crimes fall mainly into category 10 (Acts against the natural environment) (51.8%), and less often into the following Level 1 categories: 5 (Acts against property only) - 26.2%; 7 (Acts involving fraud, deception or corruption) - 11.3%; 9 (Acts against public safety and state security) - 5.4%; and 2 (Acts leading to harm or intending to cause harm to the person) - 3.6%. Single cases (0.6%) fall respectively into categories 6 (Acts involving controlled psychoactive substances or other drugs); 8 (Acts against public order, authority and the provisions of the state); and 11 (Other criminal acts not classified elsewhere).

An additional critical aspect of employing of this classification is that information about water is lost in the data collection process, with the only exception of subcategory 10012 (Acts that cause the pollution or degradation of water), which belongs to Category 10 (Acts against the natural environment). According to subcategory 10012, "Water pollution is the direct or indirect introduction of substances or energy into a body of water, water utilities or marine environment (including estuaries), resulting in harm to living resources, hazards to human health, hindrances to marine activities including fishing, impairment of the quality of sea water and reduction of amenities".

Moreover, information about the type of water (e.g. freshwater, marine water), the service stage involved (e.g. impoundment, treatment, abstraction), or the intended use (e.g. agricultural, human consumption, recreational) are not accounted for in the ICCS, not even in the disaggregating variables for the event, the victim, the perpetrator, or other data description/inclusion, making it impossible to use the ICCS as it is for analysing water crimes.

Such issues can be addressed using one of three strategies: (1) the creation of a new category in the ICCS - (12) "Acts against freshwater", with sub-categories classified according to policy relevance criteria; (2) the creation of an additional level 4 within subcategory 10012 "Acts that cause the pollution or degradation of water"; or (3) the creation of a new disaggregating variable ("freshwater" tag) within the description of the target of the offence with the following four variables: (a) state of water - SW (water quality, water service); (b) water quantity - WQ; (c) water service stage involved - WSS (abstraction, impoundment, storage, treatment, distribution, wastewater collection, wastewater treatment, other stages); and (d) intended use of water - IUW (direct human consumption, industrial, agricultural, recreational, other).

When applying the three strategies to the data collected, Strategy 3 (adding of a new tag) turned out to be the most practical, although for the time being there is still no comprehensive inventory of water-related crimes in Europe, encompassing all criminal phenomena against water. For this reason, the real or potential water crime cases (threats) collected have been framed according to the following seven main categories, which provide the basis for compiling a water crimes inventory:

1. *Water corruption* to be distinguished into so-called grand corruption, which involves political decision makers, the exploitation of natural resources, international cooperation, as well as large-scale investments and large procurement contracts; and petty corruption, involving the payment of sums of money to obtain access to water services or to avoid inspections and fines. According to the ICCS classification, corruption involves "Unlawful acts as defined in the United Nations Convention against Corruption and other national and international legal instruments against corruption" (ICCS cod. 0703);
2. *Water organised crime* refers to the activity of criminal organisations that have taken control over the management of the water or water services within a given territory. According to the ICCS classification, organised crime is defined as "Participating in the activities of an organised criminal group" (ICCS cod. 0905);
3. *Water pollution* refers to offences against the natural environment and, especially, directly against water quality. According to the ICCS classification, environmental crimes are defined as "Acts that result in the pollution of the natural environment". Cod. 10012 includes "Acts that result in water pollution or degradation" (ICCS cod. 1001 and cod. 10012);
4. *Water theft* refers to a reduction in water quantity carried out by customers of the water supply system in order to provide an economic advantage by means of a physical alteration of the water supply system. According to the ICCS classification, theft is defined as "Unlawfully taking or obtaining property with the intent to permanently withhold it from a person or organization without consent and without the use of force, threat of force or violence, coercion or deception" (ICCS cod. 0502);
5. *Water fraud* refers to fiscal artifice aimed at altering the water consumption register and achieving illicit gains. According to the ICCS classification, fraud is defined as "Obtaining money or other benefit or evading a liability through deceit or dishonesty" (ICCS cod. 0701);
6. *Water terrorism* includes the threat of terrorist attack against the water sector, water quality (e.g. by poisoning) or water availability (e.g. attacks on critical infrastructure). It may also include taking control over water services for the

illicit financing of terrorist activities. According to the ICCS classification, the crime of terrorism is defined as "Participating in the activities of a terrorist group or other individual or group acts related to terrorist offences" (ICCS cod. 0906);

7. *Water cyber-attacks* may take place via intrusions into information and technology systems; the manipulation of information or networks; or the destruction of the data of water management companies, for example through ransomware and malware attacks. According to the ICCS classification, a cyber-attack is defined as "Unauthorized access to, interception of, interference with, or misuse of computer data or computer systems" (ICCS cod. 0903).

6 Conclusions

Laws against environmental crimes address a variety of unique offences that can be categorized in different ways. The type of affected environment, such as water, can be used to categorize crimes. Many of the laws follow this format, as they pertain to a particular type of environmental media. However, some crimes involve the simultaneous violation of multiple laws. Thus, there are circumstances when it is impossible to disentangle, for instance, water pollution from soil pollution or from air pollution (e.g., the dumping of toxic chemicals can contaminate soil, off-gas into the air, and eventually leach into the watershed). Alternatively, crimes can be categorized according to the type of pollutant involved, or the disposal methods used, for instance dumping (toxic waste), mining, or spills (Gibbs & Boratto, 2017). More broadly, environmental crimes can be categorized by geography, such as domestic versus transnational environmental crime; or crimes committed across international borders (White & Heckenberg, 2014).

Nevertheless, research carried out till now has highlighted the complexities and difficulties involved in the creation of a classification of water and water-related crimes that allows for cross-country comparisons and in the adoption of standardized concepts and definitions for the systematic collection, analysis, and dissemination of data in order to fulfil the growing demand for in-depth research and analysis of these transnational crimes. Reliable crime statistics are critical for measuring changes in crime levels, monitoring state responses to crime, evaluating policies, and understanding the various facets of crime in different contexts; however, no data on water is available nowadays. At any rate, water deserves fair relevance among environmental crimes and requires a classification system like other environmental areas which have always been considered of major importance, such as illegal trade in wildlife; illegal logging and its associated timber trade; illegal, unreported, and unregulated fishing; illegal trade in controlled chemicals (including ozone-depleting substances) and illegal disposal of hazardous waste.

Against this background, the present contribution sought to carry out a general descriptive analysis of the phenomenon of water crimes in a systematic way, through the analysis of several recent case studies, in order to devise a suitable approach to the creation of a first water crimes inventory.

At the international level, UNODC ICCS (2016) can be applied to classify water and water-related crimes but presents some critical issues. To solve them, some new classification strategies have been suggested, each with its pros and cons. Trying to preserve the advantages of a statistical approach such as ICCS and, at the same time, to give to water its own "dignity" among environmental crimes by providing a framework for all crimes against water, a new approach for classifying all water and water-related crimes has been suggested, and seven main categories of water threats have been identified: water corruption, water organised crime, water pollution, water theft, water fraud, water terrorism, and water cyber-attacks.

The results of data collection show that at the EU level the most frequently reported and prosecuted crimes are water pollution and water theft, while water corruption has one of the highest risk index scores in terms of probability linked to influencing factors such as water scarcity - a rising concern in some parts of Europe - and unclear legislation (Segato et al., 2017).

After having identified the main categories of water threats to be included in the inventory, further steps in gathering useful information for the criminological analysis of the phenomenon and for international statistical purposes must be taken: the commonly recognized ways to improve water policy and fight water crimes are: (a) recognizing the extent of water misuse, abuse and crimes; (b) developing better inventories; (c) improving water-level monitoring capacity, better regulation, greater transparency, and a broadly-based external oversight of water authorities; (d) start a new systematic analysis approach, also integrated with existing data collection and exchange mechanisms by law enforcement agencies.

Although a correct classification of water and water-related crimes allows to correctly identify threats against water, it does not allow to make predictions about the outcome of possible criminal events related to the water sector. Therefore, the inventory must be considered only as a first step in tackling the complex phenomenon of crimes against water.

References

Bricknell, S. (2010). Environmental crime in Australia. Research and public policy series no. 109. Canberra: Australian Institute of Criminology. https://aic.gov.au/publications/rpp/rpp109

Brisman, A., McClanahan, W., & South, N. (2016). Water security, crime and conflict. In *Oxford handbook of criminology and criminal justice, international and comparative criminology*. Oxford: Oxford University Press.

Brisman, A., & South, N. (2019). Green criminology and environmental crimes and harms. *Sociology Compass, 13*(1), e12650.

Clarke, R. V. (1999). Hot products: understanding, anticipating and reducing demand for stolen goods. In *Policing & reducing crime: Police research series* (p. 112). London: RDS.

Eman, K., Meško, G., Dobovšek, B., & Sotlar, A. (2013). Environmental crime and green criminology in South Eastern Europe - Practice and research. *Crime, Law, and Social Change, 59*, 341–358.

ESA. (2019). *Environmental crimes.* Retrieved 23 August, 2019, from https://business.esa.int/funding/invitation-to-tender/environmental-crimes

Gibbs, C., & Boratto, R. (2017). Environmental crime. Criminology and criminal justice. *Oxford Research Encyclopaedias*. Retrieved 24 August, 2019, from http://criminology.oxfordre.com/view/10.1093/acrefore/9780190264079.001.0001/acrefore-780190264079-e-269

Gleick, P. H. (2006). Water and terrorism. *Water Policy, 8*(6), 481–503.

INTERPOL. (2012). *Environmental crime*. International Criminal Police Organization. Retrieved 23 August, 2019, from http://www.INTERPOL.int/Crime-areas/Environmental-crime/Environmental-crime

INTERPOL. (2016). *Strategic report. Environment, peace and security. A convergence of threats*. Retrieved 23 August, 2019, from https://europa.eu/capacity4dev/unep/document/strategic-report-environment-peace-and-security-convergence-threats

Johnson, H., South, N., & Walters, R. (2015). The commodification and exploitation of fresh water: Property, human rights and green criminology. *International Journal of Law, Crime and Justice, 44*(5), 146–162.

Lynch, M. J., Barrett, K. L., Stretesky, P. B., & Long, M. A. (2017). The neglect of quantitative research in green criminology and its consequences. *Critical Criminology, 25*, 183.

Lynch, M. J., Long, M. A., Stretesky, P. B., & Barrett, K. L. (2017). *Green criminology: Crime, justice, and the environment*. Oakland, CA: University of California Press.

Lynch, M. J., & Stretesky, P. B. (2013). The distribution of water-monitoring organizations across states: implications for community policing. *Policing: An International Journal of Police Strategies and Management, 36*(1), 6–26.

McClanahan, B. (2014). Green and grey: Water justice, criminalization, and resistance. *Critical Criminology, 22*(3), 403–418.

Segato, L., Mattioli, W., Capello, N., & Migliorini, M. (2017). Environmental crimes in the water sector. *European Water, 60*, 373–379.

UNEP/INTERPOL. (2016). *The rise of environmental crime. A growing threat to natural resources, peace, development and security. A UNEP/INTERPOL Rapid Response Assessment*. United Nations Environment Programme and RHIPTO Rapid Response-Norwegian Center for Global Analyses.

United Nations Office on Drugs and Crime (UNODC). (2016). *The International Classification of Crime for Statistical Purposes (ICCS)*. Retrieved 23 August, 2019, from http://www.unodc.org/documents/data-and-analysis/statistics/crime/ICCS/ICCS_English_2016_web.pdf

Water Crimes Project. (2016). *Inventory, outlook and assessment of emerging environmental crimes against water in Europe*. Retrieved 12 September, 2019, from http://www.watercrimes.eu/

White, R. D., & Heckenberg, D. (2014). *Green criminology: An introduction to the study of environmental harm*. Abingdon: Routledge.

White, R. D. (Ed.). (2009). *Environmental crime: A reader*. Cullompton: Willan Publishing.

Chapter 6
Policing Water Crimes

Gorazd Meško and Katja Eman

1 Introduction

Life originated in water, which is an essential condition for its preservation and, therefore, a source of life. Water can also be described as an environmental resource damaged by crime, the object of a crime or the means of a crime (Eman, Kuhar, & Meško, 2017). Water crime can be defined as "any punishable contravention or violation of the limits on human behaviour as imposed by national criminal legislation, which uses surface, and groundwater, or water services, as a mean for committing other crimes" (Segato, Mattioli, Capello, & Migliorini, 2017). Thus, water crimes include any intentional act that poses potential harm or damage to water. Gleick (2006) emphasises that the greatest problem stems from the fact that water resources and systems make attractive targets because there is no substitute for water. Any community, regardless of its size, will experience incredible hardship if it does not have access to freshwater either due to a physical water supply interruption, its natural scarcity or entailing contamination. Findings from the Water Crimes Project (2016)[1] reveal that water-related crimes are often recorded under other offences, such as fraud (committed by public utilities on the basis of concession agreements and others), corruption, trafficking, falsification of documents, etc.

[1] The Water Crimes Project was an international project funded with support from the European Commission (HOME/2014/ISFP/AG/EFCE/7241). It aimed at providing an inventory of water-related crimes in Europe, the first report on water crimes in Europe, and the first strategic analysis in Europe on crimes against water resources. The organisations working on the project originate from four countries: Italy, Hungary, Spain and Slovenia (Water Crimes Project, 2016).

G. Meško (✉) · K. Eman
Faculty of Criminal Justice and Security, University of Maribor, Ljubljana, Slovenia
e-mail: gorazd.mesko@fvv.uni-mb.si; katja.eman@fvv.uni-mb.si

K. Eman et al. (eds.), *Water, Governance, and Crime Issues*,
https://doi.org/10.1007/978-3-030-44798-4_6

Water has become astonishingly abundant (Crank & Jacoby, 2014) and the contamination of water resources is growing. Scarcity of clean drinking water is one of the crucial problems of the twenty-first century. The harms and risks to water supplies are associated with multiple causes, including industrial pollution, water theft and global warming. However, despite the importance of water to personal, social and economic health and wellbeing, there has been very little criminological interest in water crime issues so far (Bricknell, 2010; Brisman, McClanahan, & South, 2016; Eman, 2016; Eman et al., 2017; Eman, Kuhar, & Meško, 2016, 2018; Johnson, South, & Walters, 2016; Segato et al., 2017; White, 2003). There is an urgent need to consider the nature, dynamics and impacts of water crimes, particularly in the light of environmental drivers, such as land use, climate change and infrastructure. In order to do this, new typologies of crime have to be developed, new methodologies for research instigated and new modes of social control devised, if we are to address present and future issues adequately. White and Perrone (2015, p. 306) emphasise that questions pertaining specifically to environmental justice have been of longstanding interest and water is one of the essential items in this discussion.

To date, water crime has been poorly addressed in environmental crime and policing discourse. National and international institutions have prioritised other types of environmental crime, giving little thought to the nuances of water crimes and how they should be reflected in policing. As noted by Wright (2011, p. 333), "there has been no attempt made to draw lessons from this group of problems as a whole, nor to organize the various policy issues involved into a coherent framework for analysis". Thus, environmental crime is a high-profit/low-risk activity and crimes involving water are difficult to detect, assess, prosecute and study (Segato et al., 2017). Due to the aforementioned reasons, this chapter analyses different forms of policing environmental crime to find the most suitable approaches to detect, asses and prosecute water crimes. The authors also discuss crime prevention methods as an additional possibility to fight against water crimes.

2 Types of Water Crimes

Crimes against water are an emerging global issue and have significant impacts, although intelligence is scarce and information fragmented. Due to the lack of a systematic analytical approach leading to an agreed (working) definition and an international classification of water-related crimes, these are often recorded under other offences (e.g. fraud, corruption, trafficking, falsification of documents, organised crime, etc.). Therefore, a wide range of types of crime against water has emerged in history. Many countries and institutions have made lists of different crimes against water.

Interpol's report on environmental crime (Interpol/UNEP Environment, 2016) highlights three groups of water crimes:

1. *water fraud* (e.g. the alteration of sampling techniques or results to avoid treatment costs, thus causing negative health implications);
2. *water pollution* (i.e. the intentional contamination of water); and
3. *water theft* (i.e. the unauthorised use and consumption of water before it reaches the intended end-user).[2]

Researchers in the Water Crimes Project (2016) identified the following types of water crimes:

1. *water corruption* (includes acts similar to grand corruption, which involves political decision-makers, the exploitation of natural resources, international cooperation, as well as large-scale investments and large procurement contracts (e.g. past cases of forced water privatisation in India, Brazil, Africa and even Europe)); and petty corruption, which may be experienced daily and refers to the payment of sums of money to gain access to water services or to avoid controls and fines (The Global Initiative, 2014);
2. *water organised crime* (i.e. the activity of criminal organisations that have taken control over the management of water or water services within a particular territory. The involvement of organised crime has flooded the private sector and also penetrates many aspects of public services, such as waste management and disposal, construction and maintenance, material supply in the healthcare sector, transportation, etc. (Savona, Riccardi, & Berlusconi, 2016). Sergi and South (2016, p. 86) argue that such an involvement often results in corruption, pollution or (semi)completed projects that are inadequate or dangerous. Ruggiero and South (2010) refer to such criminal activities as "dirty collar crime".);
3. *water pollution* (i.e. offences against the natural environment and, in particular, directly against water quality. This type of water crime also includes deliberate contamination of drinking water.);
4. *water theft* (refers to acts intended to reduce the quantity of water, which are carried out by consumers in the water supply system to gain an economic advantage by physically altering the supply system);
5. *water fraud* (includes fiscal artifices aimed at adulterating registered water consumption, thus generating an illicit gain);
6. *water terrorism* (comprises activities, such as the threat of terrorist attacks on the water sector, targeting the quality (e.g. poisoning) or availability (e.g. an attack on critical infrastructure) of water (Congressional Research Service, 2010). It might also include taking control of water services to finance terrorist activities illicitly. Another scenario for a terrorist attack on domestic water supplies

[2]It is estimated that between 30% and 50% of the global water supply is illegally purchased. Regions experiencing chronic water stress (e.g. Southern Europe, Africa) and marginalised deprived areas (e.g. slums in India, Bangladesh or Brazil) are particularly vulnerable. Local communities are, therefore, forced to find alternative solutions to meet their daily water needs. In Africa, the number of unregulated wells skyrocketed from 2 million to an estimated 23–25 million in a decade. While this practice reflects poor water management, it also leads to the significant degradation of water resources (Interpol/UNEP Environment, 2016, p. 33).

involves physical attacks on water infrastructure, for instance by using conventional explosives to damage dams, pipelines or treatment plants. This could lead to the death of thousands of people, along with damage to property, damage to commercial fisheries, as well as the loss of freshwater supply and hydroelectric power); and

7. *water cyber-attacks* (which may occur by way of intrusion into an ICT (i.e. information and communication technology) system, manipulation of information or networks, data destruction, etc. of water management companies. It includes ransomware and malware attacks. Gleick (2006) also suggests that the risk of a cyber-attack on the supervisory control and data acquisition systems and networks employed by water agencies to control intake and release is not merely academic or theoretical, and thus calls for the adoption of measures aimed at reducing the probability of such water-related terrorist attacks and, should they occur, their consequences).

Barclay and Bartel (2015) studied water crimes in Australia and classified them into eight categories:

1. *water theft* (including the pumping, impoundment or diversion of water from irrigation channels, river systems, dams or groundwater bores without a licence or in contravention of licence conditions that cause changes to flows and reduce water access to neighbouring farms, livestock and riparian zone management);
2. *water contamination* (industrial contamination, contamination due to depletion of underground and surface water sources, degradation of soil, contamination of surface and groundwater by fertiliser/chemical or effluent runoff from farmland, destruction of wildlife habitat and reduction of biodiversity);
3. *waterway diversion* (referring to water theft by illegal damming of waterways, filling of tankers and deep drilled water bores);
4. *unauthorised taking of surface or groundwater*;
5. *violation of water compliance and enforcement*;
6. *water-related corruption*;
7. *water-related terrorism*; and
8. *water-related consequences of other forms of illegal or unregulated activity* (e.g. the impact of toxic chemicals and hazardous waste materials being swept up in floodwaters and polluting freshwater systems or the storage of legacy mining waste, stockpiling of tyres, repositories for chemical wastes and so on, which can be indirectly implicated in water-related crime insofar as how this occurs may violate relevant regulatory regimes and thus contribute to more significant harm than might typically be expected).

The above-presented classification clearly shows that different researchers talk about (very) similar types of water crimes. Thus, a generally accepted classification would be welcome from the policing perspective, so that police could further develop their responses in the scope of potential water crime investigations.

The uneven distribution of water around the world has led to a situation in which water theft is on the rise. Thus, the re-conceptualisation of water as a tradable commodity has increased its price and placed it in a rather prominent position in terms of potential profits it could generate for organised crime. Water theft includes the illegal acquisition of natural watercourses, as well as piped or harnessed water, both of which are described as "non-revenue water", i.e. water that is "lost" before it reaches the intended consumer. The loss of water due to theft can occur in a variety of ways, including by damaging or removing water meters and physically installing a connection to water distribution pipes, as well as stealing aid tanks (The Global Initiative, 2014). Moreover, despite the considerable interest in water, transparency in the water sector is often limited, thereby opening the door to corruption. The World Bank (2016) estimates that criminal activities and/or corruption in the water sector cause between 20% and 40% of lost revenues. Small bribes to government officials and water suppliers are used to falsify meter readings, avoid disconnections and conceal illegal connections. The growing demand for water only serves to reinforce such corrupt practices and enforce a sense of discretionary power among officials, consequently weakening a country's rule of law (Eman et al., 2017).

The final report of the Water Crimes Project (Moretti et al., 2017) revealed that deliberate contamination of drinking/groundwater, water pollution and water theft (losses in the public water supply system) were identified as the three most common water crime cases in European countries. Moreover, these cases encompassed 86% of European countries. Eman et al. (2017) warned that environmental crime groups had found an ideal business in which to make an easy profit, primarily due to the loopholes in national environmental protection legislation and the vast differences among countries in this field. This also applies to water. As long as climate change, conflict and poverty continue to aggravate the Earth's dwindling water supply, crimes against water (particularly water theft and water smuggling, together with forced water privatisation (Eman & Humar, 2017) and the violation of water quality regulations) will remain a recurring and ever more severe global problem. Gleick (2006) recommends protecting water systems through a combination of physical barriers, extensive biological and chemical monitoring and treatment, as well as the development of smart and rapid integrated response strategies. The Global Initiative (2014) emphasises that while water crimes are still viewed as petty offences or minor crimes within a national context rather than as transnational organised crime issues, it is unlikely they will ever be effectively resolved. In other words, a change of mind-set of competent agencies and organisations is needed.

3 Water Crime and Policing

3.1 Policing Environmental Crime

The story of water security can be told in terms of both historical and contemporary dimensions of the conflict. Brisman et al. (2016) point out that the pressure on water supply has increased due to population growth. The Earth's population has never been higher than today, and in many places, people are using water at unsustainable rates. This is why the police have acquired an essential role in ensuring the safety and security in the fields of the environment and water (Eman et al., 2017).

With the emergence of "new, unconventional types of crime", such as environmental crime, the police were forced to adapt to the emerging challenges of sometimes increasingly sophisticated criminal activity. Knowledge and awareness of environmental problems and dangers that may threaten "us" are the key to better preventive measures and the development of effective methods for detecting and investigating environmental crime. Environmental crime must be addressed in collaboration with all entities responsible for prevention. The training of police must enable them to carry out high-quality prevention and law enforcement. It is also necessary for police officers to become aware of the fact that the most significant problem posed by water crime is the harm it causes to people, flora and fauna. Education and training of the police constitute an essential link in protecting the environment and preventing further environmental damage. General programmes and training modules that directly contribute to a better, more effective and more efficient prevention, detection and investigation of environmental crime should be devised (Eman et al., 2017). Therefore, today's conventional policing—i.e. environmental crime policing—is somewhat lacking in success, which is why it was necessary to come up with new forms of policing environmental crime, such as problem-oriented policing, community-oriented policing, intelligence-led policing and environmental forensic science, as discussed below.

Environmental crime policing represents significant progress in the field of environmental protection, but it is also faced with several issues. Franz (2011) emphasises that the primary issues related to the enforcement of environmental crime policing arise from fragmented jurisdictions in charge of the law enforcement oversight of environmental crimes and limited resources. There is still an open question, whether the failure of local police to identify and investigate environmental crimes is due to either the lack of funding, lack of competence due unsuitable training programmes or something else (Eman et al., 2018). Moreover, policing environmental crime is similar to policing other white-collar offences, in so far as it entails vast amounts of under-policing. Water pollution that is not caused by the corporate sector is also policed ineffectively, and local law enforcement remains unable to stop small scale toxic harms, which collectively pose a huge threat to both aquatic and human life. While the clearance rate for most street crimes is considered low, the clearance rate of state environmental crime is astronomically low. Thousands of illicit acts perhaps go undetected and unabated per each detected one (Franz, 2011).

The rarity of water crimes is another problem faced by police investigating such crimes. Since they do not occur daily, police officers are not very familiar with and used to investigate water crimes. Furthermore, environmental crime may comprise ordinary water theft or water pollution cases that will occupy uniformed police officers. In contrast, other corruption- and fraud-related cases could have the characteristics of certain sophisticated forms of environmental crime for which criminal investigators will be responsible (Eman et al., 2017).

Hayman and Brack (2002) note that the prevention of further environmental crime should be extremely important. They note the wide-ranging nature and perpetrators of environmental crime, as well as the complexity of determining the illegality of proceeds. Hayman and Brack (2002, p. 4) believe that education, assistance and other incentives in the field of policing are only useful if they are complemented by a "credible threat of enforcement sanctions" against perpetrators.

With respect to effectively policing environmental crime, Schmidt (2004, p. 98) clearly states that the "precondition of effective policing is that those that enforce the law must see transnational environmental crime as a serious issue". So far, environmental crime (including water crime) has been perceived as victimless crime and characterised by the lack of awareness about its negative impacts; as a result, the police are rather slow at responding to it.

Brack (2009) and White (2009) note that prioritising transnational environmental crime is often about awareness-raising as much as anything else, which is why a concerted effort must be made to inform and educate law enforcement authorities of the importance of the task that they are charged with. Moreover, it is crucial to tie transnational environmental crime to the broader context of transnational organised crime by emphasising the links between the two phenomena. Thompson and Kanaan (2003) associate transnational environmental crime with specific priorities, such as corruption, poverty and war. It is widely agreed that greater (international) cooperation and communication is needed to combat (transnational) environmental crime. The reduction of demand and supply is crucial in decreasing the number of committed environmental crime offences. As stated by Interpol (2009, p. 6), this is "one of the greatest obstacles we must overcome" in combating environmental crime.[3]

There is, to put it simply, no easy answer to the question of where the focus of environmental crime policing should be. For example, perpetrators of environmental crime (and their *modus operandi*) differ from one region to the next, as they involve both highly organised criminal groups and individuals, whose offences include everything from massive pollution cases to ordinary animal cruelty and abuse, etc. This problem could be partly solved by increased prioritisation, training and cooperation, which would result in the increased capacity to identify the key actors in a particular region and target them specifically (Wright, 2011). Thereby, different forms of policing should be used in response to various types of environmental crime.

The literature review reveals the following forms of policing: (1) community-oriented policing, (2) problem-oriented policing, (3) intelligence-led policing, and (4) environmental forensic science, which are presented below in greater detail, since they represent possible approaches to successfully fight both environmental and water crime.

[3] For example, offences such as illegal logging, are defined by national laws where no international agreement exists. As logging legislation differs from one country to another, close communication must take place to ensure that an importing country's enforcement agencies are aware of the exporting country's laws. This is a lofty challenge, mainly as enforcement institutions are often ill-equipped and lacking in knowledge of their own legal regimes (Wright, 2011).

Community-oriented policing is considered as a crime reduction effort, which is based on collaborative relationships between law enforcement officers and the community. Moreover, it is a mechanism for reducing crime and enhancing community's satisfaction with the police. Lynch and Stretesky (2013, p. 10) note that "the idea that citizens can act as important adjuncts in the enforcement of laws and regulations has also influenced the growth of community participation in environmental policing". On the other hand, community-oriented policing does not pay enough attention to environmental issues (i.e. environmental crime). Cable and Benson (1993) believe that partnerships between environmental enforcement agencies and local communities (i.e. residents) may improve environmental quality and enforcement efficiency. Lynch and Stretesky (2013) describe such a collaboration as a situation, in which environmental agencies encourage and draw upon community members to act as participants in monitoring environmental violations—community environmental policing. Specific types of environmental crime, such as environmental pollution, illegal waste disposal or animal abuse, are happening directly in a community or are affecting a particular community. Therefore, the community can (should) draw attention to the issue(s) observed.

Problem-oriented policing (also known as POP) is an analytical method for developing strategies aimed at crime prevention and reduction. It is widely used by the police and also by some other (law enforcement) agencies. The problem-oriented policing model enables the police to systematically analyse the problems identified in a community (e.g. hot-spots), search for practical solutions and, finally, evaluate the impact of implemented changes. Problem-oriented policing also requires the police to look past traditional strategies and consider other possible approaches for addressing crime and disorder (Weisburd & Eck, 2004). Weisburd, Telep, Hinkle, and Eck (2010) describe problem-oriented policing as one of the most widely used strategies among progressive law enforcement agencies. The distinctive feature of problem-oriented policing arises from the fact that it focuses on a problem rather than on a perpetrator, which is why it is, according to Reiner (2010), also suitable for solving environmental issues. Similarly to community-oriented policing, problem-oriented policing can also be useful in responding to certain types of environmental crime that occur in a community, but are more frequent or concentrated in particular areas, such as pollution or illegal waste disposal, as well as the theft of natural resources.

The emergence of intelligence-led policing can be traced back to the first decade of the twenty-first century. It forces the police to shift from reactive to proactive policing and adopt preventive approaches to crime. The intelligence-led policing model is based on a set of ideas and principles that encourage the police to become more "strategic, future-oriented and targeted" when preventing and reducing crime (Gibbs, McGarrell, & Sullivan, 2015; Maguire & John, 2006). In other words, intelligence-led policing means that "agencies collect and analyse information about crime to create intelligence outputs that target investigations and tactical responses, aiming to disrupt, prevent and reduce crime by targeting serious offenders" (Gibbs et al., 2015, p. 254). Intelligence-led policing is consistent with problem-solving strategies based on the principles of Clarke's (1997) situational crime prevention aimed at reducing crime and victimisation. It also uses crime analysis of crime patterns to understand offenders, victims and contexts (Ratcliffe, 2008) with a view to reduce crime. Thus, past studies show that problem-solving approaches targeting specific offences reduce crime and victimisation (Clarke, 1995; Corsaro & McGarrell, 2010; Weisburd et al., 2010; Gibbs et al., 2015).

Intelligence-led policing is generally faced with implementation problems, such as the need for a significant level of resources (Wright, 2011), in the face of resource constraints (Elliott, 2012) or cross-border networks. When dealing with environmental crime, political challenges may be even greater, because they are often not managed with the same level of concern by all national agencies (Bierne & South, 2007; Gibbs, McGarrell, & Axelrod, 2010). Gibbs et al. (2015, p. 255) note, that "mistrust may be greater owing to the links between corruption and transnational environmental crime in developing countries that are often the victims of environmental exploitation". Thus, enforceable legislation and training are also often deficient. Due to poor data collection, existing information on transnational environmental criminal networks may be limited in comparison with drug cartels. Environmental enforcement agencies also lack experience in cooperating with researchers, who could support analytical efforts in the scope of intelligence-led policing. This is in contrast with police–researcher partnerships that have emerged to address other types of crime (Beal & Kerlikowske, 2010; Rojek, Smith, & Alpert, 2012). Additionally, even with the right information, in the environmental arena, procedures related to information sharing or joint efforts are often non-existent within, much less between, countries (Elliott, 2012).

In a study on the implementation of the intelligence-led policing in the scope of environmental crime, Gibbs et al. (2015) found that special attention must be paid to police procedures for the management, collection and collation, evaluation, analysis and dissemination of intelligence. Results revealed that "administrative and enforcement data can be combined to generate actionable intelligence" (Gibbs et al., 2015, p. 255) or, in other words, that it was possible to implement intelligence-led policing in addressing transnational environmental crime at the national level. Authors note that some degree of variation in legislation and training would be required to achieve the same at the international level, even though past practices in other fields, which were carried out in collaboration with Interpol, proved to be successful (e.g. Interpol's Global E-waste Crime Group (Meško & Furman, 2014)).

As already mentioned, intelligence-led policing is useful in cases of environmental crime involving corruption and fraud, in the field of organised environmental crime and in environmental crime related to terrorism or other threats.

Finally, environmental forensic science is the last form of policing environmental crime. According to Lega and Teta (2016), environmental forensic science is a forensic science that applies scientific methods to the investigation of environmental crimes and contamination incidents. The authors believe that the task of an environmental forensic scientist is to examine scenarios and actors to define the relationships in order to reveal a source, path and target, thus helping to identify the culprit of specific environmental damage. We believe that environmental forensic science can prove to be very useful in more complex environmental crime cases and should be generally used in combination with other (conventional) forms of policing environmental crime.

The prevention of environmental crime through enforcement is another possible method of responding to environmental crime. When it comes to environmental protection, the self-policing strategies follow the idea of moving away from traditional command-and-control strategies. Therefore, Stretesky (2006, p. 672) argues that self-policing is based on an assumption about the shift from traditional enforcement methods to market-based incentives. In other words, economic incentives are offered to companies to report their own environmental violations to the public authorities, thus creating "new" operational cultures (Sahramäkia, Korsellb, & Kankaanrantaa, 2015).

Sahramäkia et al. (2015) found that command-and-control strategies alone were insufficient in the prevention of environmental crime. Therefore, they suggest (high) monetary sanctions, which are sufficiently severe to deter potential offenders and change corporate behaviour. Similarly, Faure and Visser (2004, p. 63) suggested that it would be cost-effective to pursue deterrence by increasing the severity of sanctions rather than investing in enforcement.

It has also been suggested that the prevention of environmental crime through criminal sanctions could be effective because the sanctions are imposed on corporate managers who generally have a high social status and reputation that they desire to protect (du Rees, 2001, p. 111). Therefore, it has been acknowledged that the prevention of environmental crime should be extended to informal sanctions, such as naming and shaming (e.g. media attention and public shaming), which could harm the business (Faure & Visser, 2004). Nevertheless, an effective prevention of environmental crimes requires tailored sanctions and enforcement strategies in combination with a certain degree of flexibility within regulatory and enforcement regimes.

3.2 Policing Water Crime

Water crimes are a rather new trend in the field of green criminology, which have not really been addressed despite being present for quite some time. One of the reasons for such a situation arises from the narrow criminology of environmental policing. The process of criminalisation is centred on the state, which criminalises an act or event by appropriating (through the criminal justice system) the conflict between actors to itself (Hillyard & Tombs, 2007). Such an intervention is used by the state to ascribe intent to actors, sometimes using a set of dubious moral scales, particularly when dealing with environmental issues and related social harms. Kailemia (2018, p. 2161) emphasises that this "bureaucratic process individualises criminality even when it is not directly focused on an individual (i.e. 'the rich get richer, and the poor get prison')". For example, a corporation, which causes the so-called "unintentional" contamination of a freshwater resource of an indigenous community thousands of miles away (where several people die and suffer severe health consequences), gets a lower sentence (if it is sentenced at all) than a rapist or an armed attacker. This is an essential point for beginning to grapple with the ideological background against which academic discussions about environmental crimes take place and to examine how it may intersect with the analysis of specific policing methodologies. Kailemia (2018) believes that de-centring the state and its narrow definition of harm provides an opportunity for an essential shift in the field of policing ideologies related to environmental crime.

With the occurrence of water crime cases, the police are continually adapting to the challenges of investigation, because water is a "living source" that moves and changes constantly, which is why the investigation and evidence gathering can sometimes be demanding. Moreover, in terms of police work, environmental crime is often related to other types of crime; for example, theft and fraud fall in the scope of property crime investigation, while the depletion of natural resources is often related to higher economic interests and thereby to corruption and other types of white-collar or organised crime (Eman et al., 2017). In addition to ensuring cooperation between the police and other entities, while noting that their coordination should be well organised, law enforcement officers focusing on environmental crime must also have a working knowledge of various fields, from natural and social sciences to specialised expertise (Murray & Tedrow, 1992) in areas, such as chemistry, physics, biology and new technologies. White (2009) stresses that our knowledge and interest in this field might well be growing, but the more we know, the less secure we seem to be regarding the knowledge we have about environmental crime.

Even though conventional environmental policing has more than enough work with responding to water crime, we believe that the forms of policing presented above would make police work even easier:

1. *Community-oriented policing* indicates a partnership between law enforcement agencies and local communities, which is very important in the field of environmental, including water protection. In such a partnership, environmental agencies encourage community members to monitor and report environmental violations. When it comes to water crime, some types of offences, such as water contamination, water theft, unauthorised taking of surface or groundwater, as well as violation of water compliance and enforcement, occur directly in the community, thus creating an opportunity for the community to monitor the situation and report any violations.
2. *The problem-oriented policing model* enables the police to systematically analyse (environmental) problems identified in the community (e.g. hot-spots), search for effective solutions and, finally, evaluate the impact of implemented changes. Because it focuses on problems instead of on perpetrators, problem-oriented policing is, in our opinion, suitable for solving issues related to water crime. It can actually be instrumental in responding to certain types of water crimes occurring in the community, which are more frequent or concentrated in particular areas, such as water pollution, water theft, water fraud, waterway diversion (e.g. water theft through the illegal damming of waterways, filling of tankers and deep drilled water bores) and water-related consequences of other

types of illegal or unregulated activity (e.g. the impact of toxic chemicals and hazardous waste materials being swept up in floodwaters and polluting freshwater systems).

3. *The intelligence-led policing model* means that the police collect and analyse information about crime with a view of creating intelligence outputs that target investigations and tactical responses, aiming to disrupt, prevent and reduce crime by targeting serious offenders (Gibbs et al., 2015). It can be beneficial in cases of water-related corruption and fraud, water-related organised crime, water-related white-collar crime, water-related terrorism, as well as water cyber-attacks or other threats.
4. *Environmental forensic science* applies scientific methods to the investigation of environmental crimes and contamination events, where water crime is no exception. More complicated water crime cases would undoubtedly profit if they were dealt with by using forensic methods.

Different citizen groups also focus on problems related to incidents and associated environmental violations occurring in their communities. Lynch and Stretesky (2013) note that the ideas about the root causes of crime are more or less consistent with the notions of community-oriented policing. For example, citizen water-monitoring activities represent a form of social control over the threat of water pollution and foster a sense of community because people work together in their attempts to protect a stream, lake, bay or wetland located close to where they live, work or play. This is a well-known practice in the USA, where citizens developed a partnership with environmental enforcement agencies for the purpose of water monitoring.[4] Thus, Lynch and Stretesky (2013) note that in such water monitoring practices, citizens engage in informal surveillance activities that are designed to identify environmental violations and can report such violations to the authorities. O'Rourke

---

[4] *Citizen water-monitoring organisations work with state and environmental agencies by recruiting and training community members to monitor water quality, so that they can help direct state action, as well as identify sources of pollution that impact their communities. Data collected by citizen monitoring groups may even help focus enforcement efforts on specific areas.* Mauger (2002), *who studied the effectiveness of a citizens' environmental monitoring programme in Cook, Alaska, noted that "as state and federal budgets for water quality monitoring continue to decline, volunteer data will become more important in state monitoring programs" (p. 1). Over time, these water-monitoring organisations have become more formalised and have developed explicit structures, by-laws, identifiable hierarchy and a membership.*

Nevertheless, these organisations remain voluntarily. They are composed of citizen groups and, consequently, are not formal state actors. Even though these citizen groups are voluntary, it should be noted that their formation and monitoring efforts have been encouraged by federal and state environmental protection agencies, and there are many mechanisms for encouraging police-community collaboration. In many cases, the state provides training to volunteers that make up the organisation and certifies the organisations' quality assurance practices. In some ways, this takes community environmental policing well beyond traditional community policing in terms of the role of law enforcement (see Mastrofski, Worden, & Snipes, 1995). *Thus, while volunteer citizen water monitoring programme members do not possess law enforcement powers, the data they collect is collaborative and can be employed by state and federal environmental protection agencies to assess environmental quality standards identified in law* (Lynch & Stretesky, 2013, p. 12).

and Macey (2003) present a similar method used by air quality monitoring organisations called "bucket brigades", which were developed as citizen-led community policing groups that take air quality samples to help identify violations. Nevertheless, these groups resemble citizen watch programmes, which were developed as part of community-oriented policing efforts in many neighbourhoods in the 1990s (Oliver, 1998), thus complementing and helping to direct authorities' ability to "police" the environment.

Finally, water crime is a specific phenomenon. Therefore, the police must also think beyond conventional reactive methods. As a result, different types of crime prevention methods are needed to address specific water crimes (a multidisciplinary approach) allowing the application of a combination of natural and social sciences (Eman & Meško, 2014). Water crime prevention enacted by relying simply on law enforcement is not enough. It must be combined with other crime prevention strategies, such as self-regulation, smart regulation and self-policing.

4 Conclusion

Water is a natural resource that is vital for life, social and economic activities, as well as the functioning of ecosystems. It is a strategic resource for the future of the entire world. Drinking water represents a problem in many countries. The lack of water is linked to the overexploitation of natural resources and is the result of global warming (White & Heckenberg, 2014). Population growth could trigger a substantial global crisis related to water. Therefore, there is an even greater need to dramatically raise public awareness of water crime, because such acts have a considerable effect on the quality and quantity of water. As noted by Johnson et al. (2016), water should be a public good, the common heritage of people and nature, as well as a fundamental human right. Moreover, water must be provided to everyone continuously and in a quantity sufficient for personal and household purposes (Eman et al., 2017).

Water crime affects water quality, water scarcity and water insecurity. Thus, Gleick (2006) recommends protecting water systems by a combination of improved physical barriers, more extensive real-time biological and chemical monitoring and treatment, as well as by developing smart and rapid integrated response strategies. Nevertheless, policing remains the most important element of responding to water crime issues.

Policing water crime must go beyond the old-fashioned conventional policing. Otherwise, it cannot be successful enough in responding to various water crime issues. Therefore, new forms of policing are suggested, such as problem-oriented policing, community-oriented policing, intelligence-led policing and environmental forensic science. Most of them are based on problem-solving approaches and partnerships with local communities.

In general, the work of the police in this field is good and their cooperation with other agencies and organisations (even NGOs) is improving. Therefore, it is suggested that the education and training provided at the local level and to officers at individual police stations be increased. It is necessary to upgrade both general educational programmes and training in the field of environmental protection with contemporary topics related to water crime, since these currently appear to be insufficient (i.e. due to the constantly changing crime cases and the emergence of new modus operandi). Furthermore, water is a natural living resource that flows its own way. Therefore, a manual on biological and chemical treatment required in the event of water crime (i.e. water pollution or contamination) and integrated response strategies would enable the police to respond to water crime much more quickly and effectively.

The preeminent role of policing needs to be called into question and policing efforts must be tailored to meet the unique nature of water crime, while demand and supply reduction measures aiming to prevent the commission of water crimes need to be implemented and strongly supported (Wright, 2011). Moreover, prevention should not be based solely on criminal law enforcement. Sahramäkia et al. (2015) note that such enforcement efforts need to be supported by other crime prevention strategies, such as self-regulation, smart regulation and self-policing, in order for them to be effective.

In light of what we presently know about water crime and potential drivers leading to the growth of such crimes, the need for research in this area is even more pertinent. Further, Johnson et al. (2016, p. 160) suggest that green criminology and environmental protection should focus on "legal and governance frameworks that prioritise the human right to water and ecological sustainability over private interests". In addition, countries that have abundant water resources should be given a guarantee that they will not become targets of other countries or water conglomerates. On the other hand, one of the options for resolving water-related issues could be a well-thought-out arrangement allowing these countries to help people coming from areas pestered by water scarcity. In terms of water quality, it is vital to introduce constant controlled monitoring by both governments and society. Various cases throughout history demonstrate how water companies violated relevant standards in pursuit of profit, which resulted in the loss of trust and confidence among the people (Eman et al., 2017).

Finally, it is essential to raise public awareness around the world, since this is the only way to bring about major changes. Most people fail to think about water issues early enough, until, unfortunately, it often becomes too late to solve a problem because enormous damage has already been done. We must begin to protect our planet for the benefit of future generation(s), and, in doing so, we must reduce the use of chemicals and other substances that pollute water resources and the environment in general.

References

Barclay, E., & Bartel, R. (2015). Defining environmental crime: The perspective of farmers. *Journal of Rural Studies, 39*, 188–198.
Beal, P., & Kerlikowske, R. G. (2010). Action research in Buffalo and Seattle. *Police Practice and Research, 11*(2), 117–121.
Bierne, P., & South, N. (2007). *Issues in green criminology: Confronting harms against the environment, humanity, and other animals*. Cullompton: Willan.
Brack, D. (2009). Combatting international environmental crime. In R. D. White (Ed.), *Environmental crime: A reader* (pp. 485–492). Cullompton: Willan.
Bricknell, S. (2010). *Environmental crime in Australia: Research and Public Policy Series No. 109, Water Theft* (pp. 103–114). Canberra: Australian Institute of Criminology.
Brisman, A., McClanahan, W., & South, N. (2016). *Water security, crime and conflict*. New York: Oxford Handbooks Online. https://doi.org/10.1093/oxfordhb/9780199935383.013.86
Cable, S., & Benson, M. (1993). Acting locally: environmental injustice and the emergence of grass-roots environmental organizations. *Social Problems, 40*, 464–477.
Clarke, R. V. (1995). Situational crime prevention. In M. Tonry & D. P. Farrington (Eds.), *Building a safer society: Strategic approaches to crime prevention* (pp. 91–150). Chicago: University of Chicago Press.
Clarke, R. V. (1997). *Situational crime prevention: Successful case studies*. New York: Harrow & Heston.
Congressional Research Service. (2010). *Terrorism and security issues facing the water infrastructure sector.* Retrieved December 14, 2019, from https://fas.org/sgp/crs/terror/RL32189.pdf
Corsaro, N., & McGarrell, E. F. (2010). Reducing homicide risk in Indianapolis between 1997 and 2000. *Journal of Urban Health, 87*(5), 851–864.
Crank, J., & Jacoby, L. (2014). *Crime, violence, and global warming*. Abingdon: Routledge.
du Rees, H. (2001). Can criminal law protect the environment? *Journal of Scandinavian Studies in Criminology and Crime Prevention, 2*, 109–126.
Elliott, L. (2012). Fighting transnational environmental crime. *Journal of International Affairs, 66*(1), 87–104.
Eman, K. (2016). Water crimes - a contemporary (security) issue. *Crimen, 7*(1), 44–57.
Eman, K., & Humar, S. (2017). (Vsiljena) privatizacija vodnih virov – nova oblika kriminalitete zoper vodo (in ljudi) z vidika kritične ekološke kriminologije. *Revija za kriminalistiko in kriminologijo, 68*(2), 116–134.
Eman, K., Kuhar, S., & Meško, G. (2016). Providing water supply in local communities – case study of slovenia. In G. Meško & B. Lobnikar (Eds.), *Criminal justice and security in central and eastern europe. Safety, security, and social control on local communities* (pp. 367–374). Ljubljana: Faculty of Criminal Justice and Security.
Eman, K., Kuhar, S., & Meško, G. (2017). Water crimes and policing. *Varstvoslovje, 19*(2), 105–119.
Eman, K., Kuhar, S., & Meško, G. (2018). Kriminaliteta zoper vodo v Sloveniji. *Revija za kriminalistiko in kriminologijo, 69*(1), 44–53.
Eman, K., & Meško, G. (2014). *Environmental crime in criminology: Crime phenomena and development of a green criminology in Slovenia*. Saarbrücken: Scholar's Press.
Faure, M. G., & Visser, M. (2004). Law and economics of environmental crime. In H. Sjögren & G. Skogh (Eds.), *New perspectives on economic crime* (pp. 57–75). Cheltenham: Edward Elgar.
Franz, A. (2011). Crimes against water: non-enforcement of state water pollution laws. *Crime, Law, and Social Change, 56*, 27–51.
Gibbs, C., McGarrell, E. F., & Axelrod, M. (2010). Transnational white-collar crime and risk. *Criminology and Public Policy, 9*(3), 543–560.
Gibbs, C., McGarrell, E. F., & Sullivan, B. (2015). Intelligence-led policing and transnational environmental crime: A process evaluation. *European Journal of Criminology, 12*(2), 242–259.
Gleick, P. H. (2006). Water and terrorism. *Water Policy, 8*, 481–503.

Hayman, G., & Brack, D. (2002). *International environmental crime: The nature and control of environmental black markets*. London: Royal Institute of International Affairs.

Hillyard, P., & Tombs, S. (2007). From crime to social harms? *Crime, Law, and Social Change, 48*, 9–25.

Interpol. (2009). *Environmental crime programme: Strategic action plan*. Lyon: Interpol.

Interpol/UNEP Environment. (2016). *Strategic report: Environment, peace and security – A convergence of threats*. Retrieved December 15, 2019, from https://bigcatrescue.org/wp-content/uploads/2016/12/INTERPOL-UNEP-Strategic-Report-Environment-Peace-and-Security-A-Con.pdf

Johnson, H., South, N., & Walters, R. (2016). The commodification and exploitation of fresh water: Property, human rights and green criminology. International Journal of Law, *Crime and Justice, 44*, 146–162.

Kailemia, M. (2018). Problem-oriented policing of transnational environmental crimes: A social harms approach. *International Journal of Comparative and Applied Criminal Justice*. https://doi.org/10.1080/01924036.2018.1515093

Lega, M., & Teta, R. (2016). Environmental forensics: Where techniques and technologies enforce safety and security programs. *International Journal of Safety and Security Engineering, 6*(4), 709–719.

Lynch, M. J., & Stretesky, P. B. (2013). The distribution of water-monitoring organizations across states: Implications for community environmental policing and social justice. *Policing: An International Journal of Police Strategies and Management, 36*(1), 6–26.

Maguire, M., & John, T. (2006). Intelligence led policing, managerialism and community engagement: Competing priorities and the role of the National Intelligence Model in the UK. *Policing and Society, 16*(1), 67–85.

Mastrofski, S. D., Worden, R. E., Snipes, J. B., (1995). Law enforcement in a time of community policing. *Criminology 33*(4), 539–563.

Mauger, S. (2002). Effectiveness of citizens' environmental monitoring program: Exxon Valdez Oil Spill Restoration Project Final Report # 02667. Anchorage: Exxon Valdez Oil Spill Trustee Council. Retrieved December 7, 2019, from http//:www.inletkeeper.org/cemp/EffectivenessReport.pdf

Meško, G., & Furman, R. (2014). Police and prosecutorial cooperation in responding to transnational crime. In P. Reichel & J. Albanese (Eds.), *Handbook of transnational crime and justice* (pp. 323–352). Thousand Oaks, CA: Sage Publications.

Moretti, F., Segato, L., Capello, N., Mattioli, W., Monnier, C., Aleksic, D., et al. (2017). *European report on water crimes*. Szentendre: The Regional Environmental Center for Central and Eastern Europe.

Murray, R. C., & Tedrow, J. F. C. (1992). *Forensic geology*. New York: Prentice Hall.

O'Rourke, D., & Macey, G. (2003). Community environmental policing: assessing new strategies for public participation in environmental regulation. *Journal of Policy Analysis and Management, 22*, 383–414.

Oliver, W. (1998). *Community oriented policing: A systematic approach to policing*. Upper Saddle River, NJ: Prentice Hall.

Ratcliffe, J. (2008). *Intelligence-led policing*. Portland: Willan.

Reiner, R. (2010). *The politics of the police*. Oxford: Oxford University Press.

Rojek, J., Smith, H. P., & Alpert, G. P. (2012). The prevalence and characteristics of police practitioner–researcher partnerships. *Police Quarterly, 15*(3), 241–261.

Ruggiero, V., & South, N. (2010). Green criminology and dirty-collar crime. *Critical Criminology, 18*(4), 251–262.

Sahramäkia, I., Korsellb, L., & Kankaanrantaa, T. (2015). Prevention of environmental crime through enforcement—Finland and Sweden compared. *Journal of Scandinavian Studies in Criminology and Crime Prevention, 16*(1), 41–59.

Savona, E. U., Riccardi, M., & Berlusconi, G. (Eds.). (2016). *Organised crime in European businesses, (Routledge studies in crime and society)*. London: Routledge.

Schmidt, C. (2004). Environmental Crimes: Profiting at the Earth's Expense. *Environmental Health Perspectives, 112*(2), 96–103.

Segato, L., Mattioli, W., Capello, N., & Migliorini, M. (2017). Environmental crimes in the water sector. *European Water, 60*, 373–379.

Sergi, A., & South, N. (2016). Earth, water, air, and fire. Environmental crimes, mafia power and political Negligence in Calabria. In G. Antonopoulos (Ed.), *Illegal entrepreneurship, organized crime and social control: Essays in honour of professor dick hobbs* (pp. 85–100). New York: Springer.

Stretesky, P. B. (2006). Corporate self-policing and the environment. *Criminology, 44*, 671–708.

The Global Initiative. (2014). *Global initiative against transnational organised crime*. Retrieved from http://globalinitiative.net/water-smuggling/

The World Bank. (2016). *Water overview*. Retrieved December 14, 2019, from http://www.worldbank.org/en/topic/water/overview

Thompson, J., & Kanaan, R. (2003). *Final report submitted to the United States agency for international development, conflict timber: Dimensions of the problem in Asia and Africa; Volume I synthesis report*. Burlington: ARD.

Water Crimes Project. (2016). Retrieved December 7, 2019, from http://watercrimes.eu/

Weisburd, D., & Eck, J. E. (2004). What can police do to reduce crime, disorder, and fear? *The Annals of the American Academy of Political and Social Science, 593*(1), 42–65.

Weisburd, D., Telep, C. W., Hinkle, J. E., & Eck, J. C. (2010). Is problem-oriented policing effective in reducing crime and disorder? Findings from a Campbell systematic review. *Criminology and Public Policy, 9*(1), 139–172.

White, R. (2003). Environmental issues and the criminological imagination. *Theoretical Criminology, 7*(4), 483–506.

White, R. (2009). *Environmental crime: A reader*. Devon: Willan.

White, R., & Heckenberg, D. (2014). *Green criminology: An introduction to the study of environmental harm*. Abingdon: Routledge.

White, R., & Perrone, S. (2015). *Crime, criminality & criminal justice* (2nd ed.). Victoria: Oxford University Press.

Wright, G. (2011). Conceptualising and combating transnational environmental crime. *Trends in Organized Crime, 14*, 332–346. https://doi.org/10.1007/s12117-011-9130-4

Part II
Water Protection and Water Crimes: Case Studies

Chapter 7
Water Legal Protection in the Canary Islands: A Brief Description

Luis Javier Capote-Pérez

1 Introduction

Water constitutes one of the principal aspects of environmental protection. As a resource, it is the example of the necessity of a logical control of all natural resources and of a sustainable development. Maybe 75% of our planet is water, but only a minimum percentage of this is potable or usable for human consumption. In a context where concerns about future are increasing, it is a logical consequence that, in the legal perspective, the access to that resource could be conceived as a human right. Menéndez Rexach (2010) mentions the water right in the group of other essential rights as health, social services or environment.

Water Law and the right to use water are aspects of the modern legal regulation of that good, conceived as a resource and as a product of that resource (Orozco Muñoz, 2016) and are pivotal aspects of the increasing interest about water management and regulation in Environmental Law. The existence of a legislation relative to those areas is, certainly, not new, but the evolution of its norms reflects the growing concern about environment protection and sustainable development. The case of the Canary Islands is a good example, where geological, geographical and historical circumstances have traditionally affected the regulation of the access to water resources.

Chapter is made under researching Project DER2017-83970-P, 2018-2020 "La nueva información registral: requisitos, eficacia y aplicación práctica"

L. J. Capote-Pérez (✉)
Department of Basic Legal Disciplines, Universidad de La Laguna, La Laguna, Spain
e-mail: lcapote@ull.es

K. Eman et al. (eds.), *Water, Governance, and Crime Issues*,
https://doi.org/10.1007/978-3-030-44798-4_7

2 A Brief Description of the Canary Islands

According to Custodio and Cabrera (2002) the Canary Islands are one of the Autonomous Regions of the Spanish State and is constituted by two provinces over seven major islands—eight, according to the 2018 Autonomy Statute—each one with its own insular government and administration—*Cabildo*. The archipelago is on the west of the Saharan coast of Africa, at roughly 100 km of closest distance, and its islands are spread along 500 km. The region is between 27° 31′ N – 29° 30′ N latitudes and 13° 30′ W – 18° 10′ W longitudes. Its area is about 7,446 square kilometres and has a population of approximately 2,120,000 inhabitants, living specially in Gran Canaria and Tenerife. The density is really high, with almost 475 inhabitant /square kilometre in Gran Canaria.

The islands are of volcanic nature and origin, by accumulation of effusive materials over the ocean floor, with correspondent intrusive structures. The sub-oceanic part of the structure and also the sub-marine basement arose in some islands after their tectonic raising and posterior erosion or shift of the sub-aerial part. The beginning of their formation—in the actual incarnation—is late-cretaceous and first sub-aerial effusions are from the Miocene period.

The archipelago's location in the Saharan area of high pressure supposes that pluviometry at sea level is really very limited, particularly in the eastern islands. The average at that level is 50–100 mm/year and it increases lightly with the altitude. However, in the higher islands—actually in all of them except for Fuerteventura and Lanzarote—there is an interesting orographic effect over wet trade winds—*alisios*—from the northeast, that produces abundant rainfall on the northeast faced hillsides, going beyond 900 mm/year in some places. Similar effects happen with occasionally stormy west and southwest winds that sometimes could bring abundant rain falls to normally drier regions.

The main consequence of this phenomenon is that in limited rainy areas important freshwater resources are generated, and, sometimes, surface runoffs due to the general high permeability of soils. Much of the rain goes to recharge and recover volcanic aquifers. Subterranean freshwater flows to coast but can emerge before if there are lithological or topographical changes. Details about this process are different in each island, due to the orographic differences between them.

The conclusion in this point is that water in the Canary Islands is a scarce and valuable resource. Its demand has been growing from the period previous to the Castilian conquest—when aboriginal island population was not really large and, consequently, hydric needs were not high—to the present time—when the number of inhabitants and tourism as the principal economic activity have notably and alarmingly increased the demand and exploitation of this resource. This fact is reflected in the legal regulation of the access, use and protection of water in the Canaries legal system.

3 A Brief Description of Water Regulation in Spanish Law

According to Menéndez Rexach (2010) actual conception of Spanish Water Law begins with 1866 Water Act (Ley de Aguas de 1866) whose Articles 31 and 33 establish the nature of public domain of water streams—Art. 31 for rainfall waters in torrents and wades and Art. 33 for water in rivers and creeks-. Historical circumstances of the Codification of Civil Law in Spain determined that many regulations that should be incorporated to Spanish Civil Code (1889) were collected in special acts. When, in 1889, the Civil Code enters, at last, into force, this conjunctural legislative strategy became permanent and references to Water Law in that legal text coexisted supplementary with the next special regulation, 1879 Water Act (1879). More precisely, only Articles 407 and 408 of Spanish Civil Code introduce express mentions to water regulation:

Article 407
The following waters are of public domain:

1. *Rivers and the natural courses thereof;*
2. *Continuous or discontinuous waters of springs and streams flowing in the natural courses thereof, and the riverbeds;*
3. *Waters which spring in a continuous or discontinuous manner in land which is of public domain;*
4. *Lakes and lagoons created by nature in public land, and the rivulets thereof;*
5. *Rainwater which flows on cliffs or watercourses, if the course is also of public domain;*
6. *Underground waters in public land;*
7. *Water found in areas when public engineering works are taking place, even if performed by a concessionaire;*
8. *Waters which spring in a continuous or discontinuous manner from the private property of individuals, of the State, the province or villages, from the time they exit such properties;*
9. *Any excess from fountains, drains and public establishments.*

Article 408
The following waters are private goods:

1. *Continuous or discontinuous waters which spring from privately owned plots of land, whilst they remain in such land;*
2. *Lakes and lagoons and the rivulets thereof, created by nature in such plots of land;*
3. *Underground waters located in such plots of land;*
4. *Rainwater which falls therein, whilst it does not exit the boundaries thereof;*
5. *The beds of continuous or discontinuous running waters formed by rainwater, and those of any streams which flow through land and properties which are not public domain.*

In any irrigation channel or aqueduct, the water, the bed, the boxes and the banks shall be deemed an integral part of the plot of land or the building for which the waters are destined. The owners of the plots of land through which or through whose boundaries the aqueduct were to pass may not allege ownership thereof, nor any right to use its bed or banks, unless it is based on deeds of ownership which evidence the right or ownership claimed thereby,

Rules contained in Spanish Civil Code coincide with previous regulations of Water Acts and both establish different status for the resource, depending on the public or private condition of land where it is, but introducing the primary principle that rivers and natural courses are part of public domain. Water has, therefore, a legal nature of special property—Orozco Muñoz (2016) and Moreu Carbonell (2011)—and this condition is extended from the resource (water in its natural state) to the product (taken for its use).

In 1978 Spanish Constitution (Constitución Española, 1978) water resources are defined as part of the public domain:

Article 132.

1. *The legal system governing public domain and community property shall be regulated by law, on the principle that they shall be inalienable and imprescriptible and not subject to attachment or encumbrance.*
2. *The property of the State public domain shall be that established by law and shall, in any case, include coastal area, beaches, territorial waters and natural resources of the economic zone and the continental shelf.*
3. *The State and National Heritage, as well as their administration, protection and preservation, shall be regulated by law.*

According to Valencia Martín (2017) this constitutional rule transforms all waters in the hydrological cycle into public domain. The need of a reasonable use of natural resources were basis to establish this legal status, whom regulation was developed in 1985 Water Act (Ley 29/1985, de 2 de agosto, de Aguas). This change was not considered as an *ope legis* expropriation, according to Constitutional Court resolutions—STC 227/1988, 29 November (ECLI:ES:TC:1988:227)—because the new regulation admitted the possibility of maintaining some previous private rights under special circumstances (Fernández Farreres, 2018).

Nowadays, Spanish Water Law still combines aspects of Public and Private Law. Moreu Carbonell (2011) remarks this particular regulation remembering that, according to the 1985 Water Act—now 2001 Water Act (Real Decreto Legislativo 1, 2001)—the regulation of this special property is primarily regulated by Administrative Law and Civil Code precepts are supplementary. In accordance with Article 2 of Water Act, the hydraulic public domain is integrated with continental above-ground and below-ground waters, courses of rivers, creeks and torrents and lakes', lagoons' and reservoirs' beds. Therefore, the public condition contains not only water as a resource, but also physical elements which contain it or along it runs and the product of processes of desalination. Certain private ownership rights are maintained with more or less freedom than in the old regulation of 1866 and 1879

Water Acts. In any case, public and private properties are submitted to the regulation of use limits of hydraulic public domain and to EU Environmental Law.

More than 30 years after the 1985 Water Act entered into force, the current scenario of Spanish Water Law, as Moreu Ballonga (2011) says, reflects an uncompleted and distorted conversion into the public domain of the resource. The original idea was the consideration of all water as part of one hydrological cycle, but the introduction of transitory rules which established the general principle of non-retroactivity respect of previous private rights, changed practically that premise. Owners had the election of maintaining their rights according to the former regulation—but with the mentioned new limits—or inscribing them in a special water register system, transforming them into administrative concessions, 50 years after registration. In 2010, only between 10% and 20% of private water owners in Spain had registered their rights.

The conclusion in this point is that in Spanish Law, the property of water is only nominally part of the public domain, because the general rule of conversion into that public condition is accompanied by exceptions that, practically, suppose the conservation of private domain over a great percentage of water resources. This situation occurs specially in the Canary Islands, where historical motives cause an augmented version of the national panorama.

4 The Case of the Canary Islands

Sánchez Jordán (2000) keeps that the great number of specialities in water regulation in the Canary Islands permits the affirmation of the existence of a Canary Water Law. Origins of that peculiarity begin with the incorporation of the Archipelago to the Crown of Castilla, in the sixteenth century. Those singularities are connected, firstly, with the way of sharing the property of water immediately after the conquest and, after that, with an intense participation of private initiative in activities of delivery, exploitation and distribution of the resource. The evolution of water ownership created endemic institutions such as *comunidades de aguas* (water communities) and *heredamientos* (water inheriting communities) and, until the sixties, there was no doubt about private condition of water in the region.

In the 1985 Water Act a rule was established, according to stipulations of the State Water Law would have special effects in Canary Islands, according to the constitutional distribution of competences between Spanish State and Spanish regions. The mentioned Constitutional Court resolution STC 227/1988, 29 November established that 1985 Water Act, a State Act, was applicable to set the status of public domain of water resources, also in island regions as the Canary or Balearic Islands, recognizing the competence of the Canary Autonomous Region to legislate about above-ground and below-ground waters, but never in contradiction of state rules. According to this competence distribution, Canary Water Law would repeat, at its regional level, the problem of relations between water resources as public domain and the respect of previous private rights over them. The classic dichotomy acquires in this region a special dimension, based in the historical origins of water utilization.

The conquest and later incorporation of the Canary Islands to Castilla has two periods. In the first one, Lanzarote, Fuerteventura, La Gomera and El Hierro were conquered by feudal lords whose descendants transferred their rights to the Crown of Castilla *(islas de señorío).* In the second one, Gran Canaria, La Palma and Tenerife were conquered directly by the kingdom and incorporated to the royal heritage *(islas de realengo).* During the period after the conquering, lands were distributed with rights to use water for irrigating, but not with ownership nature. There was, already in the sixteenth century, a genuine preoccupation about a rational use of water and, according to some chronicles, destinations of water were, firstly, the population and, secondly, industrial and agricultural uses. However, the introduction of the cultivation of the sugar cane supposed a dissociation, a "divorce" between land and water. The resource was converted into an independent object of commerce. Inheritors of primitive rights of use transformed them into ownership rights and increased them, consolidating the private condition of water, through the institution of the water inheriting community *(heredamiento).* These communities are part of the resource exploitation during the following centuries. In the nineteenth century, the recognition as private of some water resources, according to the aforementioned 1866 and 1879 Water Acts and Civil Code, reinforced their ownership. On the other hand, the growing demand of the liquid element was produced by the introduction of new and successive exploitations as cochineal, banana and tomato. Therefore, another institution, the water community, arose with the objective of finding underground waters and distributing them between communards. By means of the 1956 Water Inheriting Communities of the Canary Islands Act (Ley de 27 de diciembre de 1956), the Spanish State recognized legal personality to water communities, assuming their crucial role in the islands' economy.

As stated above, under the rule of the 1978 Spanish Constitution, the Canary Islands acquired the condition of Autonomous Region with competences of its own. According to the constitutional distribution, interpreted in the Constitutional Court resolution STC 227/1988, 29 November, State water rules were applicable in the Canary Islands but regional particularities introduced the possibility and the need of a regional water act. The process of elaboration of Canary legal rules reflected one again previous tensions between the intention of transforming all water resources into public domain and former private rights over them. The first Autonomous Canary Island Water Law—1987 Canary Water Act—established a stricter regulation than 1985 State Water Law. So, in this way, if State Law, as mentioned above, gave owners the choice of maintaining their rights according the former regulation or inscribing them in a special water register system, transforming them into administrative concessions, 50 years after registration, Canary Law reduced the period of decision from 3 years to two, establishing an automatic transformation from private ownership to administrative concession, which would have only an inextensible duration of fifteen years. In both legal texts, the election of maintaining the former regulation was the conservation of the private ownership, so it is not actually clear why this regional act created the great opposition between water owners that it provoked. Anyway, its entering into force was successively delayed until its substitution by another regional rule, the 1990 Canary Water Act (Ley 12 de Aguas de la Comunidad Autónoma de Canarias, 1990), currently applicable and much more respectful and protective with ownership rights and endemic communities. Again, there is a choice to inscribe private rights or not to do so, with the consequences of the transformation into a temporary right of use and exploitation or the maintenance of the already existing status.

The conclusion in this point is that the problem of water regulation in the Canary Islands is the same as in the rest of Spain, but with peculiarities as the "divorce" between water and land in the process of distribution, use and transmission of rights over those resources and the lack of the liquid element and its over-exploitation, especially during the last 150 years. The original desire and legal intention of converting all components of the hydrological cycle into public domain is in contrast with regulations that allow the permanency of private ownership over water.

5 Criminal Law Protection of Water in the Spanish Law

According to the State organisation established by the 1978 Spanish Constitution, Autonomous Regions do not have any competence about Criminal Law, according to Article 149.1.VI. Only the State holds the exclusive competence over this matter. Therefore, Penal Law protection of water in Spanish Law is to be found in the Spanish Criminal Code.

The Criminal Code (1995) reflects the growing preoccupation about environmental protection, initiated in Spanish Law in the 1978 Constitution (Muñoz Conde, López Peregrín, and García Álvarez, 2013):

Article 45

1. *Everyone has the right to enjoy an environment suitable for personal development, as well as the duty to preserve it.*
2. *The public authorities shall safeguard rational use of all natural resources with a view to protecting and improving the quality of life and preserving and restoring the environment, by relying on essential collective solidarity.*
3. *Criminal or, where applicable, administrative sanctions, as well as the obligation to make good the damage, shall be imposed, under the terms established by the law, against those who violate the provisions contained in the previous clause.*

This constitutional precept introduces the idea of natural heritage as an object of a bifrontal right/duty of enjoyment and preservation. Within it concepts such as suitable environment and natural resources are included, with the commandment of conservation, restoration and public punishment and reparation of damages. Water resources are included in this constitutional precept and, consequently, the basis to use Criminal Law rules as instruments for its protection and establish one Environmental Criminal Law.

Martos Núñez (2006) defines Environmental Criminal Law as the set of rules that regulates violations against the environment and their juridical consequences. The legal protection of the environment as a matter of public interest is the legally protected right or asset and comprehends a wide concept of environment, that includes the quality of life, natural resources, flora and fauna protection, the spatial planning and historical heritage conservation—in this point, *vid.* also Capote Pérez (2018)-.

Within Environmental Criminal Law, Blanco Lozano (2000) distinguishes and specifies the existence of a Water Criminal Law, justifying the penal intervention in the value of the protected asset and the severity of possible violations. Water Criminal Law has a bi-dimensional condition, as part of both juridical branches.

In the Spanish Criminal Code (1995), we can find precepts that concretize the penal protection of the environment and, more precisely, water, under the concept of *felonies against natural resources and the environment,* in Articles 325–331 Criminal Code (1995). We are going to focus on three precepts, with express mentions to water offences: Article 325, about emissions and spillages; Article 326, about waste shipments; Article 326 BIS, about crimes against the balance of natural systems. If we look at the original wording of Article 325, we can perceive the achievement of general rules about Environmental and Water Criminal Law:

Whoever, breaking the laws or other provisions of a general nature that protect the environment, directly or indirectly causes or makes emissions, spillages, radiation, extractions or excavations, filling with earth, noises, vibration, injections or deposits, in the atmosphere, the ground, the subsoil or the surface water, ground water or sea water, including the high seas, even those affecting cross border spaces, as well as the water catchment basins, that may seriously damage the balance of the natural systems shall be punished with a sentence of imprisonment from 2 to 5 years, a fine from eight to 24 months and with special barring from his profession or trade for a period from 1 to 3 years. Should there be risk of serious damages to the health of persons, the sentence of imprisonment shall be imposed in its upper half.

On the one hand, we can see that water legal protection is embodied through penal rules in the protection of the already mentioned wide concept of environment. On the other hand, the care about the resource is based not in an effective damage, but in the possibility of that, as we can perceive in the expression "may seriously damage". The precept describes a great number of acts, all defined by the nature of being legal infractions- the consequences of which could severely affect not only continental water but also sea water. The same structure can be perceived in Articles 326 and 326 BIS where the potential production of a serious damage is included in the criminalization. In the Judgement of the Appellate Court of Santa Cruz de Tenerife (Sixth Section) no. 23/2017, 26 January 2017 (ROJ: SAP TF 939/2017 - ECLI: ES:APTF:2017:939) felonies against natural resources and the environment are defined as complex structures with the premise of an illegal act and the production of a severe damaging fact—an emission, a spillage—where the seriousness of the risk over the environment establishes the limit between administrative and criminal offences. The implementation is going to be analysed through two Judgements of the Appellate Court of Santa Cruz de Tenerife, selected as examples.

The first case, contained in the Judgment of the Appellate Court of Santa Cruz de Tenerife (Second Section) no. 710/2010, 18 March 2010 (ROJ: SAP TF 1758/2010 - ECLI:ES:APTF:2010:1758) concerns the acts of a landowner in the town of Tacoronte (Tenerife). During almost 15 years, he used one part of his property as an illegal garbage dump, exploiting it economically, receiving payments made by everyone who wanted to abandon all sorts of waste—organic, mechanical, chemical… He disobeyed every advice or prohibition delivered by competent Public Administrations and infringed many environmental rules. Court condemned him in application of former Articles 325 and 326, basing the decision in the fact that the felony included the creation of a serious risk of damage derived from spillages provoked by the accumulation of rubbish. In the reconstruction of events, the abandonment of car batteries is present, liquids could eventually be filtered in a process of leachate and contaminate subsoil water. It is not a fact but a logical evaluation of a risky and possible scenario.

The second case, contained in the Judgment of the Appellate Court of Santa Cruz de Tenerife (Second Section) No. 82/2017, 20 February 2017 (ROJ: SAP TF 93/2017—ECLI: ES:APTF:2017:93), concerns the existence of a town dump in the Municipality of Puntagorda (La Palma) and the possibility of felonies described in Articles 325 y 326 and allegedly responsibility of two former majors of its Council. During the proceeding, the analysis of possibilities of a severe risk of damage to the environment was a central point and, more precisely, water resources were object of a specific consideration and research. Studies concluded that the presence of the dump did not suppose a risk for subsoil water, because of geographic and hydrologic reasons. The principal aquifer of La Palma was in the high part of the island, and Puntagorda is in a volcano cone in the oldest sector. The flow of water to the sea in this region, usual due to effects of gravity, is minimal, due to the volcanic nature of the soil. Water wells closer to the dump area are unproductive and are not being used anymore. The conclusion was that there was no potential risk for aquifers or subsoil water.

The conclusion in this point is that, despite the fact that there is no Canary Criminal Law and State Criminal Code is applicable in the Autonomous Region, geographical and hydrological particularities of the Archipelago have strong influence in the application of rules about felonies against water, conceiving the protection of this resources in the penal perspective as part of a wide concept of environment.

6 Conclusions

In a more and more populated world, water is one of the most valued resources. This affirmation acquires different particularities depending on the country or the region, but it is a principle generally admitted that our quality of life and, actually, our survival, depends on a wise management of natural resources and environment. In the achievement of this objective, Law has an important contribution to make.

In Spanish Law, and especially since the 1978 Spanish Constitution, the protection of the environment, considered in a wide perspective as natural heritage, is one of the governing principles of economic and social policy. Constitutional commandments justify and reinforce the initiated process of conversion into public domain of water resources, but not without tensions with ownership rights over a great part of them. The story of the old dichotomy between public and private property of water has a particular chapter in the Canary Islands, where vicissitudes of the incorporation to Castilla and of resources' distribution has made that, in many ways, the public condition of water is not really effective. In spite of legal restrictions applicable to private owners, their property is, in a large proportion, still actually private.

However, the private condition of that proportion of water resources is not a limit to protect all with Criminal Law rules. Water is part of a wide concept of environment, conceived as the legally protected asset and its defence, through punishment of felonies described in the Spanish Criminal Code is not based on the fact of a severe damage but on the risk of its production. Over the old conception of ownership as the right to use, to exploit and to abuse *(ius utendi, fruendi et abutendi)* there is a social function limit of private property, in this concrete case the exigence of protection of natural heritage.

Current regulations of water are a good example of private ownership restrictions, although there are practical difficulties for their effective application. In a region as the Canary Islands, where water resources are equally scarce and precious, the need of a real protection and, overall, of a full awareness is more acute. The application of criminal rules about felonies against the environment is the best evidence that there are still too many things to do in this area.

References

Blanco Lozano, C. (2000). In J. M. Bosch (Ed.), *La tutela del agua a través del Derecho penal*. Barcelona.

Canary Islands Act (Ley de 27 de sobre heredamientos de aguas del archipiélago canario). (1956). Retrieved 23 July, 2019, from https://www.boe.es/datos/pdfs/BOE//1956/365/A08223-08225.pdf

Canary Water Act (Ley 12 de Aguas de la Comunidad Autónoma de Canarias). (1990). Retrieved 23 July, 2019, from https://www.boe.es/eli/es-cn/l/1990/07/26/12/con

Capote Pérez, L. J. (2018). Bancos de protección de la naturaleza y Registro de la Propiedad. *Revista Crítica de Derecho Inmobiliario, 768*, 1753–1788.

Criminal Code. (1995). Retrieved 24 July, 2019, from https://www.boe.es/eli/es/lo/1995/11/23/10/con

Custodio, E., & Cabrera, M. C. (2002). Cómo convivir con la escasez de agua. *Boletín Geológico y Minero, 113*(3), 243–258.

Fernández Farreres, G. (2018). Artículo 132. In A. Rodríguez-Piñero, N. Bravo-Ferrer, & M. Casas Baamonde (Eds.), *Comentarios a la Constitución Española. XL Aniversario. Tomo II* (pp. 862–875). Madrid: Boletín Oficial del Estado. Ministerio de Justicia. Tribunal Constitucional. Fundación Wolters Kluwer España.

Martos Núñez, J. A. (2006). Introducción al derecho penal ambiental. In J. A. Martos Núñez (Ed.), *Derecho penal ambiental* (pp. 17–19). Madrid: ExLibris Ediciones.

Menéndez Rexach, A. (2010). El derecho del agua en la legislación española. In J. Agudo González (Ed.), *El Derecho de aguas en clave Europea* (pp. 25–70). Madrid: La Ley-Wolkers Kluwer España.

Moreu Ballonga, J. L. (2011). Artículo 408. Aguas de dominio privado. In A. Cañizares Laso, P. De Pablo Contreras, J. Orduña Moreno, & R. Valpuesta Fernández (Eds.), *Código Civil Comentado. Volumen I. Título preliminar – De las normas jurídicas, su aplicación y eficacia. Libro I – De las personas. Libro II – De los bienes de la propiedad y de sus modificaciones (Arts. 1 a 608)* (pp. 1630–1640). Cizur Menor: Civitas – Thomson Reuters.

Moreu Carbonell, E. (2011). Artículo 407. Aguas de dominio público. In A. Cañizares Laso, P. De Pablo Contreras, J. Orduña Moreno, & R. Valpuesta Fernández (Eds.), *Código civil comentado. Volumen I. Título preliminar – De las normas jurídicas, su aplicación y eficacia. Libro I – De las personas. Libro II – De los bienes de la propiedad y de sus modificaciones (Arts. 1 a 608)* (pp. 1625–1629). Cizur Menor: Civitas – Thomson Reuters.

Muñoz Conde, F., López Peregrín, C., & García Álvarez, P. (2013). *Manual de derecho penal medioambiental*. Valencia: Tirant Lo Blanch.

Orozco Muñoz, M. (2016). Agua subterránea: situaciones y relaciones jurídicas. *Anales de la Facultad de Derecho, 33*, 9–38.

Sánchez Jordán, M. E. (2000). La titularidad (y el aprovechamiento) de las aguas en Canarias. In R. Bercovitz & J. Martínez Simancas (Eds.), *Derechos civiles de España* (pp. 4021–4056). Pamplona: Aranzadi.

Spanish Civil Code. (1889). Retrieved 23 July, 2019, from https://boe.es/buscar/pdf/1889/BOE-A-1889-4763-consolidado.pdf

Spanish Constitution (Constitución Española). (1978). Retrieved 24 July, 2019, from https://www.boe.es/eli/es/c/1978/12/27/(1)/con

Valencia Martín, G. (2017). *Jurisprudencia constitucional y medio ambiente*. Cizur Menor: Thomson Reuters Aranzadi.

Water Act (Ley 29/1985 de Aguas). (1985). Retrieved 23 July, 2019, from https://www.boe.es/eli/es/l/1985/08/02/29

Water Act (Ley de Aguas de 1866). (1866). Retrieved 23 July, 2019, from https://www.boe.es/datos/pdfs/BOE//1866/219/A00001-00004.pdf

Water Act (Ley de Aguas de 1879). (1879). Retrieved 24 July, 2019, from https://www.boe.es/datos/pdfs/BOE//1879/170/A00799-00805.pdf

Water Act (Real Decreto Legislativo 1). (2001). Retrieved 23 July, 2019, from https://www.boe.es/eli/es/rdlg/2001/07/20/1/con

Chapter 8
Water Crimes in Cyprus

Ioanna Hadjiyianni, Andreas Kapardis, and Nicos Pavlides

1 Introduction

The planet's resources, including water resources, are being put under increasing pressure by human activities as the Earth's population continues to grow. Pollution is a recent phenomenon in the history of our planet and a human problem with devastating effects. Over time, various human activities have been contaminating water bodies-lakes, rivers, aquifers, and groundwater- and water pollution is the leading cause of death and disease worldwide due to water-borne diseases, for example (West, 2006). Common causes of water pollution are sewage, nutrients (e.g., chemicals) used in agriculture, chemical waste, radioactive waste, oil pollution, plastics, and thermal pollution. The legislature in Cyprus has paid particular attention to reducing water pollution. Before focusing on environmental law on water quality and management, let us first consider criminological research and the environment in Cyprus.

I. Hadjiyianni (✉) · A. Kapardis
Department of Law, University of Cyprus, Nicosia, Cyprus
e-mail: hadjiyianni.ioanna@ucy.ac.cy; kapardis@ucy.ac.cy

N. Pavlides
NAP Regulatory Compliance Services Ltd., Nicosia, Cyprus
e-mail: pavlidesn@napcompliance.com

K. Eman et al. (eds.), *Water, Governance, and Crime Issues*,
https://doi.org/10.1007/978-3-030-44798-4_8

2 Criminological Research and the Environment in Cyprus

Criminological research in Cyprus is a comparatively recent development and so is the interest in white-collar crime (Sutherland, 1940) while the interest in crimes against the environment (UN Environment Programme, 2016) is even more recent. Even though some work was carried out, inter alia, into juvenile delinquency and prisoner recidivism in the 1980s (Kapardis, 1983, 1986), it was not until 1999 that an international was organised on economic crimes in Cyprus (Kapardis, 2001). Even though interest by criminologists in the protection of the environment from human interference is not new, including in Greece (Alexiades, 1981), the first international conference on crimes against the environment did not take place until the end of 2016 (Kapardis, 2018) and there, for the first time, the audience was introduced to Green Criminology (Andersen, 2014). The sad reality is that research into environmental crimes is almost non-existent in Cyprus. This is of particular concern in view of the fact that, as the Global Risks Perception Survey 2014 highlighted, two of the top 10 risks in terms of impact over the next ten years are environmental risks: firstly water crises and failure to adapt to climate-change (World Economic Forum, Global Risks, 2015).

In order to set the scene for what follows, first a few words about the climate that has prevailed since the Turkish invasion of Cyprus in 1974 and the occupation of its northern part that resulted in 200,000 refugees being forced to move to the south of the island. During the period 1974-early 1990s, as Theopemptou (2018, p. 153) points out: "During these years, there wasn't any noticeable care or protection for the environment. That was the period environmental protection groups appeared first on the island. Some of the most active and vocal environmental NGOs were the Friends of Akamas and the Ecological Movement. Political parties of the left also started to show interest in environmental issues. Because of the collective action of these groups, some protection was gained".

Cyprus became a full member of the European Union (EU) in 2004, and environmental related legislation was transposed into the national legal order. However, both the concern about the environment as well as knowledge regarding environmental issues and threats to the environment still characterises most politicians and decision-makers alike. The current 'crime debate' in Cyprus centres around drugs, juvenile anti-social behaviour, and the occasional allegation of corruption. Not surprisingly, therefore, the investigation, prosecution and punishment for crimes against the environment is relatively a recent phenomenon, and environmental injustice continues unabated by economic elites and politicians in the name of economic development and profitability. However, not all is bleak on the environmental front on Aphrodite's island, and attention next turns to legislation for the protection of the environment in Cyprus.

2.1 Legislation for the Protection of the Environment in Cyprus

Generally speaking, efforts have been made by the relevant authorities in Cyprus to implement EU Directives on the environment (Pyrgou, 2018; Theopemptou, 2018). This includes: Law 22 (1)/2012 for the Protection of the Environment Through the Criminal law that Brought Cyprus in line with the EU Directive on the protection of the environment through criminal law (2008) and provides for a fine up to 500,000 euros and imprisonment up 10 years or both and that profits from offences against the environment are to be dealt with as illegal proceeds from crime.

Cyprus can be said to have some additional good legislation:

1. Law 189 (1)/2007 concerning the responsibility for the prevention and restoration of damage to the environment, inter alia, provides:
   - A deadline for the recovery of the restoration cost.
   - Administrative fine up to 315,600 euros and appeal possible to the relevant Minister.
   - Offences can be committed by a corporation.
   - Non-compliance is punishable with a fine up to 315,600 euros.
2. Law 106/1/2002 concerning water pollution (Amended the 1991 law)
   - Provides for a fine up to 79,000 euros and/or 3 years imprisonment.
   - The Council of Ministers appoints an inspector to enforce the law who can impose a fine up to 316 euros.
   - This law abolished the court order suspending the operation of a convicted company.

However, cases such the outcome of particular environmental studies (e.g., Limassol Marina), that of the Salt Lake near the old Larnaca Airport and river pollution involving a known Cypriot multinational company, inter alia, provide evidence that: (1) very few cases involving environmental offences are prosecuted; (2) penalties imposed are but a slap on the wrist; and (3) corruption in some government departments and in the private sector underpins such practices. The fact of the matter is that having good legislation is not enough because:

- political will is vital for effective law enforcement.
- adequately resourced services are required.
- corruption negates effective law enforcement.

The following two sections analyse key legislative instruments stemming from the EU on the protection of the environment through criminal law, particularly relating to water quality, and their implementation and enforcement in Cyprus. While EU law has made significant steps towards the protection of water resources, including the criminalisation of certain acts, considerable discretion is still left to the Member States for the implementation and enforcement of EU Directives. As demonstrated below through the discussion of relevant cases before local courts, in relation to Cyprus, this discretion has mainly meant that environmental protection legislation has not been effectively enforced and most water crimes go unpunished.

3 Environmental Law on Water Quality and Management: The Case of Water Crimes and Cyprus

> Water quality is a profoundly important issue not only because it raises questions about environmental quality but also because water is a necessity for life. (Fisher, Lange, & Scotford, 2013, p. 552)

Environmental law has sought to regulate water quality as a dynamic issue given that water moves through the natural environment and works in a cycle (Fisher et al., 2013). Water pollution can be addressed in many different ways and at different levels of governance. Through permits for industrial installations issued by local authorities, through goals for achieving a 'good water status' in different bodies of water, through international conventions for the protection of shared watercourses that promote cooperation between countries. One important and emerging mode of regulating environmental pollution, including water pollution is through criminal law. This is not a widespread mode of addressing water pollution, at least at the international level, but it is increasingly emerging at the regional (EU) and national levels. This chapter does not seek to analyse all relevant environmental legislation, but rather focuses on relevant legislation that seeks to criminalise infringements of water quality and management legislation.

The criminalisation of environmental law violations has emerged primarily for improving compliance. This is because criminal offences are thought to carry 'a social disapproval of a qualitatively different nature compared to administrative or compensation mechanisms in civil law' (Preamble of Directive, 2008/99/EC; Mitsilegas, Fitzmaurice, & Fasoli, 2016). At the same time, reliance on criminal law is not necessarily always a good thing as it requires investigation by the police triggered by monitoring efforts or a complaint (European Report on Water Crimes, 2019). Also, the use of criminal law does not always lead to more effective compliance due to the weakness of national systems, reliance on the initiative of the public prosecutor, a high threshold of proof and the length of proceedings (EFFACE Synthesis, 2016).

3.1 The International Level

Many different international treaties have emerged with the aim of protecting water resources. These include the codification of the customary rule of international law to protect the marine environment as a whole, including in the high seas, under the UN Convention on the Law of the Sea (1982), regional seas agreements such as the Barcelona Convention on the protection of the marine environment and the coastal region of the Mediterranean (Convention for the Protection of the Mediterranean Sea Against Pollution, 1976) and the UN Conventions on international watercourses (UNECE Convention on the Protection and Use of Transboundary Watercourses and International Lakes 1992; UN Convention on the law of the Non-Navigational Uses of International Watercourses, 1997).[1] However, such international agreements do not usually recognise the need for the creation of criminal offences.[2] Instead, they promote cooperation among states for the management of surface and groundwater bodies that cross state boundaries.

One rare example of an international, yet mostly regional, a treaty which directly addressed the criminalisation of the protection of the environment is the Council of Europe Convention on the Protection of the Environment through Criminal Law.[3] This Convention includes different kinds of water crimes, including the intentional offence of 'unlawful operation of a plant in which a dangerous activity is carried out and which causes or is likely to cause … substantial damage to the quality of air, soil, water, animals or plants'. The Convention categorises intentional and negligent offences. It also specifies that sanctions shall include imprisonment and pecuniary sanctions and may include reinstatement of the environment. In many respects, it goes further than any EU legislation on the matter by categorising the different offences and setting out procedural and substantive elements of criminalisation. While the Convention opened for signature in November 1998, it has never entered into force due to insufficient ratifications.[4] This is mainly due to political reasons, including the focus in this region on the EU's action in approximating the national laws of the Member States (Vagliasindi, 2017).

[1] Note that Cyprus has not signed either Convention.

[2] One exception of an international treaty which can be interpreted as accommodating criminal offences it the MARPOL Convention, which is discussed below.

[3] Note that Cyprus has neither signed nor ratified this Convention.

[4] Only one out of the three required ratifications have taken place (Estonia).

3.2 The EU Level: EU Criminal Environmental Law and Water Crimes

The criminalisation of violations of EU environmental legislation is an area where we have seen the EU exercising its competences to limit the freedom of the Member States, which for many years possessed considerable discretion in deciding how to control infringements of EU environmental legislation. This change has mainly occurred through the adoption of the Environmental Crime Directive (ECD, 2008) in 2008, which requires Member States to establish criminal offences for violations of certain instruments of the EU environmental acquis and national legislation implementing it (ECD, 2008). Previously, EU Member States were free to determine penalties for infringement of EU legislation, with EU law merely requiring that these penalties be 'effective, proportionate and dissuasive'.[5] EU criminal environmental law has essentially come to supplement pre-existing environmental legislation to harmonise the approaches of the different member states and strengthen the enforcement of EU environmental legislation through criminal law.

The ECD was initially adopted as a framework decision under the third pillar on EU justice and home affairs. Following a challenge before the Court of Justice of the European Union ('CJEU'), this decision was annulled as the Court ruled that the appropriate legal basis was rather the one relating to the environmental policy under the first pillar.[6] The CJEU clarified that the EU could take measures which relate to the criminal law of the Member States when necessary to render environmental provisions fully effective (C-176/03 para 48). A similar approach was adopted in relation to the framework decision strengthening the criminal law framework for the enforcement of the law against ship-source pollution. The CJEU clarified the limits of this competence, which does not extend to the determination of the type and level of the criminal penalties to be applied.[7] The CJEU therefore approved the use of criminal law as a 'means to an end' (Mitsilegas et al., 2016), that 'end' being the better enforcement of EU environmental legislation, while preserving some discretion for the Member States to determine the exact level of criminal sanctions to be applied within their national legal orders.

Against this background, the ECD requires Member States to attach criminal sanctions to violations of EU environmental legislation or provisions of national legislation that implement it. In particular, Member States have to ensure that, at least,[8] *the conduct listed in Article 3 of the Directive is deemed as a criminal offence when unlawful and committed intentionally or at least with serious negligence.*

---

[5] For example, see Directive, 2010/75/EU of the European Parliament and of the Council of 24 November 2010 on industrial emissions (integrated pollution prevention and control) [2010] OJ L 334/17, Article 79 discussed below.

[6] Article 175 EC (now article 192 TFEU) C-176/03 *Commission v Council* EU:C:2005:542.

[7] Case C-440/05 *Commission v Council* EU:C:2007:625, para 70

[8] Member States are in principle free to criminalize additional conduct in accordance with Article 193 TFEU.

Article 3 specifies three types of water offences. First, the discharge, emission or introduction of a quantity of materials or ionising radiation into air, soil or water, which causes or is likely to cause death or serious injury to any person or substantial damage to the quality of air, the quality of soil or the quality of water, or to animals or plants (ECD, 2008, Article 3a). Second, the operation of a plant in which a dangerous activity is carried out or in which dangerous substances or preparations are stored or used and which, outside the plant, causes or is likely to cause death or serious injury to any person or substantial damage to the quality of air, the quality of soil or the quality of water, or to animals or plants (Article 3d). Third, the production, processing, handling, use, holding, storage, transport, import, export or disposal of nuclear materials or other hazardous radioactive substances which causes or is likely to cause death or serious injury to any person or substantial damage to the quality of air, the quality of soil or the quality of water, or to animals or plants (Article 3e). What constitutes 'unlawful' is determined by reference to a list of relevant EU legislation set out in Annex A of the ECD, which includes the Water Framework Directive (Directive, 2000/60/EC establishing a Framework for Community Action in the field of Water Policy), further discussed below, and other water-related legislation.

Under the ECD, inciting, aiding and abetting the intentional conduct of Article 3 is also considered a criminal offence (Article 4). Additionally, the Directive establishes the liability of legal persons when this conduct is carried out for their benefit (Article 6), thereby properly allocating responsibility to polluters, given that corporations often play a crucial role in the commitment of environmental crimes (Farmer, Faure, & Vagliasindi, 2017, p. 48). However, legal persons do not have to be subject to *criminal* sanctions, but rather the Member States have to ensure that they are punishable by 'effective, proportionate and dissuasive penalties' (Article 7).

Despite these significant steps at EU level, the ECD's structure and the use of imprecise terms may lead to legal uncertainty. As per the CJEU's case law, the ECD does not set the exact level of criminal penalties, which are left to be decided by the Member States but have to be effective proportionate and dissuasive (Article 5). As the EU Directive does not set the exact level of punishment, implementation in the Member States differs considerably, ranging from 6 months to life in prison and/or fines ranging from 200,000 to 600,000 euros. In this respect, when Cyprus eventually implemented the ECD,[9] criminal liability was established somewhere in the middle of the range, setting up to 10 years imprisonment and/or 500,000 euros fine for the most serious offences (Law 22(I)2012 on the protection of the environment through criminal law, 2012, Article 4). Beyond this flexibility provided to the Member States, some terms used in establishing the offences such as 'substantial damage' and 'serious negligence' are not clearly defined in the ECD. This is in contrast to the Council of Europe Convention, which defines the criminal offences covered in more direct ways. Due to the delicate balance of division of competences

[9]The Commission referred Cyprus to the CJEU on 22 March 2012 for failure to transpose the Directive into national law (the deadline for transposition was December 2010). See http://europa.eu/rapid/press-release_IP-12-296_en.htm?locale=en

between the EU and the Member States, the exact delimitation of the offences is determined by national legislation implementing the Directive. For example, the Cypriot implementing law defines serious negligence and damage.[10] Nonetheless, these definitions can be interpreted in different ways, which makes the role of the national courts crucial in the implementation and enforcement of these criminal offences.

Because the main EU criminal law instrument criminalises violations of other EU environmental legislation, it is necessary to briefly explain the operation of key EU directives that regulate water quality and management, on the basis of which criminalisation occurs. The key legislative instrument on water policy at EU level is the Water Framework Directive ('WFD'), which sets legally binding targets for water quality and quantity of surface and groundwater bodies including coastal waters. Unlawful violations of this legislation should carry criminal sanctions with the application of the ECD (ECD, 2008, Annex A).

Under the WFD, Member States have an obligation to 'protect, enhance and restore' all bodies of water 'with the aim to achieve' good water status by 2015 (Water Framework Directive, Articles 4(1)(a)(ii) and 4(1)(b)(ii)). In relation to surface water, 'water status' is determined in reference to the ecological status of the body of water, relating to the quality of the structure and functioning of associated aquatic ecosystems, and to the chemical status of the body of water—relating to the concentrations of pollutants that should not exceed specific environmental quality standards. The status of surface water is classified as 'high', 'good', 'moderate', 'poor' or 'bad'. In relation to groundwater, water status is determined in relation to the quantitative and the chemical (qualitative) status and is classified as good or bad. It already becomes obvious that the concept of good water status is complex and uncertain, demonstrating the difficulties of governing water quality through regulation.

The Member States are also under a non-deterioration obligation, to take the necessary measures to maintain the quality of a body of water (Water Framework Directive, Articles 4(1)(a)(i) and 4(1)(b)(i)). This has been interpreted by the CJEU as a legally binding obligation, also binding the Member States when approving individual projects. The CJEU clarified that *'the Member States are required—unless a derogation is granted—to refuse authorisation for an individual project where it may cause a deterioration of the status of a body of surface water or where it jeopardises the attainment of good surface water status or of good ecological potential and good surface water chemical status by the date laid down by the directive.'*[11] This non-deterioration obligation has an autonomous ranking, and it is not merely an instrument at the service of the obligation to enhance the status of bodies of water (Paragraph 49).

---

[10] Law, 2012(I) 22 defines serious negligence as an 'act and/or omission which leads to non-compliance with a level of conduct in relation to a risk that is reasonably perceived' and damage as the direct or indirect adverse change or degradation of air, soil and / or water, animals and / or plants and / or their habitats.

[11] C-461/13 *Bund für Umwelt v Bundesrepublik Deutschland* EU:C:2015:433 para 51.

Despite these strict obligations on the Member States, the WFD also incorporates considerable discretion as to its implementation by the individual Member States in the form of exceptions/derogations. First, it is possible for the Member States to designate some water bodies as 'artificial or heavily modified' by human activity so that they cannot achieve natural conditions and therefore the obligation on the Member States is to achieve good ecological potential rather than status. This allows human benefits from existing water uses to be taken into account so that the costs of improving the quality of these water bodies are not disproportionate and unduly expensive (Water Framework Directive, Article 4(3)). Second, there is the possibility to extend beyond the 2015 deadline. This extension to 2027 is possible if the Member State concludes that the necessary improvements cannot 'reasonably' be achieved on time for reasons of 'technical feasibility', 'disproportionate expense' or because natural conditions do not allow timely improvement in the status of the body of water (Article 4(4)). These two exceptions are the ones mostly used in Cyprus' River Basin Management Plans.[12] Additional exceptions include less stringent objectives than good status because it would be infeasible or disproportionately expensive to achieve the objectives of the Directive (Article 4(5)).

The legal framework established by the WFD consists of different steps of implementation, starting from the Member States classifying different bodies of water, which provides the baseline for improvement and monitoring. The Member States are required to produce a river basin management plan that should be updated every six years, and which includes the general description of characteristics, maps and boundaries of water bodies as well as the assessment of their status and any exceptions applied in accordance with Article 4 (Article 13). It essentially sets out how the objectives will be met within a particular timeframe. Within the RBMPs, Member States are required to set out a programme of measures, which include prior authorisations of projects, prohibitions on abstractions and discharges. In terms of implementation, the WFD provides that 'Member States shall determine penalties applicable to breaches of the national provisions adopted pursuant to this Directive. The penalties thus provided for shall be effective, proportionate and dissuasive (Article 23).' Following the ECD, violations of the WFD should carry criminal sanctions, the level of which is determined by the Member States.

Beyond the Water Framework Directive, supporting directives such as the Bathing Water Directive (2006b), the Drinking Water Directive (1998) and the Groundwater Directive (2006a) are also included in the list of environmental legislation to which criminal sanctions should be attached at the national level in accordance with the ECD. Furthermore, the ECD criminalises violations of the Urban Waste Water Treatment Directive (1991), which imposes obligations on the Member States in relation to the collection, treatment and discharge of urban wastewater.

At EU level, criminalisation of water offences has also emerged specifically in relation to ship-source pollution. This is a distinct example where the EU has taken action to implement a multilateral environmental agreement through criminal sanc-

[12] www.moa.gov.cy/moa/WDD/wfd.nsf. Accessed 31 May 2019.

tions by harmonising EU law. As mentioned at the beginning of this chapter/section, not many multilateral agreements on water pollution impose criminal sanctions. In relation to water pollution from ships, however, the International Convention for the Prevention of Pollution from Ships (MARPOL), while not explicitly indicating criminal penalties, can be interpreted to allow the use of such sanctions (1973). In particular, it prohibits discharges of oil and noxious substances and provides that sanctions shall be established in accordance with the law of the state under whose authority the ship is operating. These penalties 'shall be adequate in severity in order to discourage violations and shall be equally severe irrespective of where the violations occurred' (Article 4). With Directive, 2009/123/EC (2009), the EU requires the Member States to attach criminal sanctions to violations of Directive 2005/35/EC on ship-source pollution. Similarly to the ECD, the Directive on ship-source pollution does not set the exact level of sanctions but rather requires that criminal sanctions are effective, proportionate and dissuasive.

Arguably, the EU's implementation of MARPOL through criminal sanctions goes beyond the international requirements by requiring ship-source discharges of polluting substances to be regarded as infringements if committed with intent, recklessly or by serious negligence (Article 4). In contrast, MARPOL indicates infringement when all reasonable precautions have been taken, and the owner or the master of the ship acted either with intent to cause damage, or recklessly and *with the knowledge* that damage would probably result. The EU Directive does not include 'with knowledge' and covers the liability of people other than the master or owner of the ship. While this approach was challenged before the CJEU, the Court refused to examine the compatibility of the EU Directive with MARPOL because the EU was found not to be bound by it.[13] Arguably, the CJEU's approach was unduly restrictive and it should have instead sought to reconcile the interpretation of 'serious negligence' under the Directive with the term 'with knowledge' under MARPOL so as to demonstrate respect of international law and to provide clarity to the implementation of the Directive and MARPOL by the Member States (Mitsilegas et al., 2016). This discussion duly demonstrates the challenges of criminalising environmental law infringements at different levels of governance, given that criminal law is an area which challenges state sovereignty and the delimitation of competences between the EU and the Member States is very complex.

Apart from EU legislation specifically attaching criminal sanctions to violations of environmental legislation, there are additional instruments which regulate water pollution and to which some Member States, including Cyprus, have attached criminal sanctions. These include the Industrial Emissions Directive (2010), which provides a significant mode of regulating environmental pollution by large industrial installations through the requirement of a permit by the competent national authority. This permit addresses the overall environmental performance of the industrial installation, covering, among others, emissions to air, soil and water, waste genera-

[13] Case C-380/06 *Intertanko and others v Secretary of State for Transport* EU:C:2008:312. Note that Cyprus was one of the states to submit observations supporting that the Directive went beyond international law.

tion, energy efficiency and prevention of accidents. Pollution under the Directive is defined broadly as 'direct or indirect introduction, because of human activity, of substances, vibrations, heat or noise into air, water or land, which may be harmful to human health or the quality of the environment, result in damage to material property, or impair or interfere with amenities and other legitimate uses of the environment' (Article 3(2)). The requirement for permits essentially guarantees that pollutant inputs will respect the maximum limits set by legislation. According to EU law, the Member States are free to determine the penalties applicable to infringements of the national provisions implementing the Directive and such penalties shall be 'effective, proportionate and dissuasive' (Article 79). Cyprus opted to apply criminal sanctions to infringements of the national implementing law. In particular, in accordance with the Industrial Emissions (Integrated Pollution Prevention and Control) Law of 2013 (Law 184 (I)/2013), infringement is considered a criminal offence and leads to imprisonment and/or fine.

The final piece of legislation at EU level worth mentioning in the context of water crimes is the Environmental Liability Directive (2004). The Directive implements the 'polluter pays principle' by requiring operators to take preventive action for imminent environmental damage and to bear the costs of clean-up measures to remedy environmental damage. This includes obligations on water damage defined as 'any damage that significantly adversely affects:

1. the ecological, chemical or quantitative status or the ecological potential, as defined in Directive, 2000/60/EC, of the waters concerned, with the exception of adverse effects where Article 4(7) of that Directive applies; or
2. the environmental status of the marine waters concerned, as defined in Directive, 2008/56/EC, in so far as particular aspects of the environmental status of the marine environment are not already addressed through Directive, 2000/60/EC' (Article 2(1)(b)).

The WFD, discussed above, thus plays an important role in the implementation of the Environmental Liability Directive. The Environmental Liability Directive establishes different kinds of liability for different kinds of damage. For non-dangerous activities, other than those listed in Annex III, liability is fault-based. For dangerous activities, listed in Annex III, it establishes strict liability. In relation to water pollution, dangerous activities include the 'discharge or injection of pollutants into surface water or groundwater which require a permit, authorisation or registration in pursuance of Directive, 2000/60/EC' (Annex III (5)). A causal link between the activity carried out, and the damage is always required and has proved to be one of the major challenges for the effective operation of the Directive. Importantly the Environmental Liability Directive does not establish tortious claims between individuals but rather requires the competent authority to ensure its implementation. However, any affected natural and legal persons and environmental NGOs have the right to request competent authority to take remedial action (Article 12). While not an instrument of criminal law, the provisions of the Environmental Liability Directive can help prevent environmental crime by making perpetrators liable for the consequences of their action and clean-up measures (Fisher et al., 2013).

Overall, the above discussion demonstrates that significant steps have been taken to address water pollution through criminal law at EU level and the combination of multiple directives which carry criminal offences bears significant promises for the better enforcement of environmental protection legislation. However, the extent to which criminalisation of environmental protection infringements leads to better results for the environment and those affected by violations largely depends on the functioning of the internal legal order of each Member State and the willingness and resources of prosecutors and courts to enforce the relevant legislation in practice.

4 Enforcement of Environmental Protection Legislation in Cyprus

> When I visit a country, I do not examine if it has good laws, but whether the existent laws are effectively enforced, because good laws exist everywhere.
> Montesquieu, '*The Spirit of Laws*'

As analysed earlier in this chapter, the Republic of Cyprus, having acceded to the EU in 2004, has transposed all relevant EU environmental protection directives into Cypriot law. A significant caveat needs to be placed here: approximately 37% of the area of the Republic of Cyprus has been occupied by Turkey since 1974, following the Turkish invasion. The occupied part of Cyprus includes about 60% of the coastline of the island. The application of the European *acquis* has been suspended in relation to the occupied areas that are not under the control of the internationally recognised government of the Republic of Cyprus. There is no accurate information on environmental matters in the occupied areas, and therefore, discussion in this section of the chapter is limited to the free part of Cyprus. The continuing Turkish occupation presents by itself an immense environmental challenge and significantly limits the ability and scope of applying and enforcing the laws and regulations for the protection of the environment in the Republic of Cyprus. Further, the economic effects of the Turkish invasion and occupation of part of the island in 1974 were devastating: one-third of the population of the island was displaced having lost their homes and all real estate property in the occupied north of the island and were housed in hastily constructed refugee housing in the south. Approximately 60% of the GDP generating components of the Cypriot economy was lost as a result of the invasion and occupation.

In light of the above, at a time when the concepts of sustainable development were emerging, and the international norms towards the protection of the environment were being formed, e.g. in the Declaration of the United Nations Conference on the Human Environment in 1972,[14] the Republic of Cyprus had its attention

[14] Second preamble of the Declaration of the United Nations Conference on the Human Environment in 1972: '*The protection and improvement of the human environment is a major issue which affects the well-being of peoples and economic development throughout the world; it is the urgent desire of the peoples of the whole world and the duty of all Governments.*'

turned to the dire consequences of the war and occupation, and its efforts were focused on reconstruction and re-starting the economy. The environment in the non-occupied south part of the island also fell a victim of the war and continuing to-date occupation of the north. In the name of re-starting the economy, which was heavily reliant pre-1974 on tourism, agriculture and light industry, the setting and enforcing environmental standards was seldom, if ever, significant consideration of the State and responsible authorities. Permits for hotel construction were readily granted in the 1980s without adequate planning, civil engineering and environmental impact studies; the seaside hotels were allowed to channel their sewage waste into the sea; factories and industries were allowed to dispose or bury their pollutant waste with no concern for the environment and the water resources. One particularly telling case is that of *Askarel*, as described by the former Commissioner of the Environment of the island Mr. Charalambos Theopemptou (2018, p. 154): Askarel is a highly toxic and carcinogenic PCB chemical used for industrial insulating/coolant purposes. In 1986 a scrap metal company used transformers to dismantle and sell the scrap metal, containing about 160 tonnes of this chemical. About 30,000 cubic metres of soil up to 20 m in depth were affected at the time by more than 50 tonnes of Askarel. At the time of the incident, the solution provided by the relevant Ministry was to bury the remaining Askarel underground in two watertight tanks. From time-to-time, the matter resurfaced in the news. In 2016, Mr. Theopemptou, in his capacity as Member of Parliament, asked the relevant Minister whether the authorities were considering removing the chemicals from underground and have it processed either in Cyprus or abroad. The reply from the minister stated that they were not considering any action. As time passes by, the risk of Askarel leaking into the groundwater increases as the sealant materials age increases. No penalties or prosecutions are known to have taken place in relation to this matter. No priority was given to disarm this time bomb which threatens the water resources of the island.

The philosophy underlying harmful interventions to the environment in Cyprus after 1974 and throughout the 1980s, as highlighted in the above case, was that harmful interventions to the environment were either not considered a crime, or were considered, at most, a victimless one, and certainly not a priority for prosecution. The environment was not a legal person, and unless a person could decisively prove the link of causality between human interventions to the environment and damage to health or personal property, the authorities would usually not prosecute or they would take measures with great delay and tolerance, and individuals would have no standing to bring a case against anyone.

The cases brought to court and led to the successful conviction in relation to water crimes and environmental pollution, in general, are very few and the penalties imposed have generally been inadequate to achieve discouragement of future offences. The main reasons are the following:

1. *The low priority given by authorities to protect the environment through criminal law and/or high tolerance exhibited*: The Environmental Crime Directive, 2008/99/EC was transposed into national legislation only in 2012, with Law 22(I)/2012 on the protection of the environment through criminal law. This is the

most effective legislative weapon in the Republic of Cyprus in relation to the protection of the environment through criminal law, setting, as mentioned earlier in this chapter, up to 10 years of imprisonment and/or €500,000 fine for the most serious offences. Seven years have passed since the enactment of this law, and not a single case was brought before the courts to-date for violations. Further, prior to the enactment of this law, relevant legislation provided for very light sentences, grossly inadequate for specific or general discouragement. Even within the pre-existing framework of light sentences, the courts, as noted by Pyrgou (Pyrgou, 2018, p. 148) are not yet prepared to impose harsh fines in the field of environmental harm.

2. *Non-independence and lack of enforcement powers of the Office of the Commissioner of the Environment*: The Office of the Commissioner of the Environment in Cyprus was not given the status of an independent authority with supervisory and enforcement powers, but its mandate is to act as an advisory organ to the State and make recommendations to other state organs, e.g. ministries for improvements. It cannot enforce the laws or impose any administrative fines. It has been given very limited resources.
3. *There is no Ministry of Environment in Cyprus, which can examine horizontally, supervise and enforce the environmental legislation across all government and private sector; and this contributes to a fragmented and not-infrequently inefficient dealing of the matter*: The main State organ with supervisory powers in relation to the environmental, is the Department of Environment under the Ministry of Agriculture. Its role includes receiving applications for licenses which it can accept or reject. Cases of disposal of end-of-life vehicles or metals causing pollution or posing a risk to public health (a frequently observed issue in Cyprus) may also be examined by the Road Transport Department under the Ministry of Interior. The Ombudsperson also has the power to examine cases of complaints against government departments in relation to the environment and provide opinions. However, as in the case of the Commissioner of the Environment, the Ombudsperson has no enforcement powers, and the scope of the Ombudsperson's work is limited to the examination of potential violations of government and state organs, and not of the private sector. The Auditor General of the Republic of Cyprus may also examine potential cases of violations of state organs where there are elements of misappropriation or maladministration of public funds, including municipalities, but it can only furnish opinions; if a criminal offence appears to have taken place, he will send the brief to the Office of the Attorney General of the Republic, which has the exclusive responsibility for the prosecution of criminal offences in the Republic of Cyprus.
4. *Lack of sufficient expertise of the investigative organs and the prosecution authority (the Office of the Attorney General of the Republic), in relation to the investigation and prosecution of offences against the environment; and of the courts in relation to the adjudication of such cases.*

The following criminal cases illustrate the above.

1. *Attorney General of the Republic of Cyprus v. Vasiliades and Rainbow Bleaching and Dyeing Company Ltd* (2001) 2 SCD 65 (the '*Rainbow Bleaching*' case)—*tolerance by authorities exhibited over long periods of time.*

   In this case, the defendants admitted the disposing of treated water waste without having the necessary license. This was a breach of the Water Pollution Control Act of 1991. The offence lasted for four years. They had a prior conviction of a similar offence. The first instance court imposed a fine of 1000 Cyprus pounds (about € 1700) to the director of the factory and 1500 Cyprus pounds (about € 2550) on the company. At first instance, a second, more serious offence of actual pollution of the water was rejected by the court for insufficiency of evidence presented. The Attorney General appealed the first instance decision, seeking an order for closure of the factory in addition to the fines, until and unless a permit was issued. The Supreme Court accepted the appeal of the Attorney General and ordered the closure of the factory. The facts of the case reveal that the authorities had tolerated the continuing operation of the factory for years, and encouraged the defendants to apply time and again for the necessary license. *The defendants' applications were rejected six times (!), but no measures had been taken for the closure of the factory at an earlier stage.*
2. *Director of the Department of Environment of the Ministry of Agriculture* v. *Nicolaides*, case no. 5214/14, decision date 11.4.2016—*authorities' delays in bringing a case, election by prosecution to bring the matter to a lower court which had limited powers for imposing sentences, sentence finally imposed materially inadequate on the basis of the gravamen of the conduct of the defendant.*

   The defendant, an owner of a vehicles' repair business, admitted breaching several provisions of the Waste Disposal Law of 2011. *The facts of the case reveal that the defendant, with contumelious disregard for the environment and public health, threw waste (end of life vehicles, buses, tires, metal and plastic waste) alongside the main road, and the affected area stretched for 5 kilometres.* The Road Transport Department issued a fine of € 1000 when the problem was initially detected in 2011, and in 2012 sent him a letter asking him to remove all waste and deliver them to licensed waste disposal businesses, as well as to restore all areas that had been affected by his actions and negligence. The case against him was filed when the defendant failed to pay the fine and comply with the instructions of the Road Transport Department. *Three and a half years later, the defendant had yet to comply with the instructions of the Department.* The decision states that:

   > the defendant's conduct is characterized, with any degree of leniency, completely unconscionable, created an unprecedented pollution site with evident danger of infections…the defendant, with only his economic interest in mind and being essentially indifferent to the damaging effects to the environment and public health, managed to turn nature to a wasteland of metal and rubber.

   The law provided for a maximum sentence of € 500,000 and/or 3 years imprisonment. *The prosecuting authority had, however, elected to bring the matter to a first instance court, whose powers were limited to the imposition of fine up to € 80,000. The court imposed a suspended sentence of 6 months imprisonment and a fine of merely € 10,000.*

3. *Attorney General of the Republic v. Renos Stavrides*, case no. 8650/2013, decision date 26.9.2017—*lack of sufficient expertise in investigating and prosecuting environmental offences, fragmentation of relevant departments*

   The defendant was indicated with 19 counts for violations of the Water Pollution Control Act of 2012, as an accomplice with Plyntex Public Ltd. ('Plyntex'), and having aided or abetted Plyntex for the disposal of liquid waste to a stream that ended to a salt lake in a coastal city of Cyprus. The indictment stated that the offence occurred on or about 26th October 2011. The first witness was an officer of the Department of Environment of the Ministry of Agriculture. He stated that Plyntex operated an industrial facility for washing garments, and the defendant was its director. On 26th October 2011, following a complaint from the health inspector of the relevant municipality, he inspected the site and saw that Plyntex had placed a plastic pipeline from its facilities, which could direct liquid waste to the stream that ended to the salt lake. He presented relevant photographs of the site. During cross-examination, he, however, stated that the photographs neither showed any actual waste disposal through the pipeline nor did he see liquid flowing from the pipeline during his on-site inspection on the date that the indictment stated that the offence took place (26th October 2011). The second witness for the prosecution was the health inspector of the relevant municipality, and she also stated that on the relevant date she did not observe the flowing of waste or sewage liquids from the pipeline starting from Plyntex's facilities. The testimonies of the first two witnesses for the prosecution also conflicted between them on a crucial point: the first witness stated that the defendant, who was present during the on-site inspection, admitted to the offence, whereas the second witness stated that the defendant merely said that he was aware of the waste disposal, but this was not done pursuant to his instructions. A third witness for the prosecution who was also present at the on-site inspection further differentiated the evidence in relation to what the defendant said during the inspection: the witness stated that the defendant told them that he had placed the pipeline but had made arrangements with the Sewage Board of the city for the disposal of the waste to the sewage system—not to the salt lake. The fourth witness was the Head of the State General Laboratory, which had contacted tests on samples of liquid waste taken from the stream and of samples taken from Plyntex's tanks. She gave evidence that the samples she tested could cause pollution. However, under cross-examination, she said that she did not know whether the two samples she collected, of liquid from the stream and the tanks of the company matched. She said that with the tests she conducted, she was unable to reach this conclusion. Additional, more specialised tests would have been needed in order to be able to reach this conclusion, which had not taken place.

   The Court, faced with so many inconsistencies of the evidence offered in relation to the on-site inspection, the fact that the witnesses could not confirm that there was actual disposal of waste to the stream on the date stated on the indictment and relying heavily on the evidence of the Head of the State General Laboratory, found that the Prosecution failed to prove its case and acquitted the defendant of all 19 counts.

4. *Attorney General v. Bouygues Batiment International and others* Criminal appeals 130/2011 and 163/2011, decision of the Cyprus Supreme Court of 23.06.2015—Fragmentation of relevant departments, lack of sufficient expertise of the investigative organs and the prosecution authority (the Office of the Attorney General of the Republic), in relation to the investigation and prosecution of offences against the environment; and of the courts in relation to the adjudication of such cases.

   The defendant is a French construction company with a global presence,[15] which was the main contractor for the construction of the new international airport in Larnaka, a coastal city of Cyprus. A whistle-blower, a mechanical engineer who was working for the Company, filed a written complaint to the Department of Environment of the Ministry of Agriculture, that the defendant company during the work undertaken in 2009, instead of proceeding to examine offers for the cleaning of waste of the company, disposed the waste through a pipeline from the engine room of the airport to an open tank, which channelled the waste in the salt lake adjacent to the airport.

   Seven persons were charged: Bouygues (the main contractor), two sub-contractor companies and four natural persons who were in managerial/director positions in the companies. The indictment contained 12 counts. Seven counts involved the disposal of toxic liquid waste and related offences, in violation of the Water Pollution Control Act of 2012. Three counts related to the disposal of waste in a public place without a license, in contravention of the Solid and Dangerous Waste Management Act of 2002. Almost all persons were acquitted at first instance in 2011, and the Attorney General appealed against the acquittal.

   The lack of experience and, dare to say, low priority given to the prosecution of this case was evident from the very start. The charges against three out of the four natural persons were dropped since the prosecution proved unable to serve them the indictment abroad. This was despite the fact that the company is headquartered in the EU, it is a major global player, and prides itself as a leader in sustainable construction. Assistance by the French authorities could have been obtained, and the directors and officers of a company with such a financial position and visibility could not just vanish.

[15] The website of the Group of the company presents the profile of the Group as: 'Bouygues Construction is a global player in construction, with operations in more than 60 countries. It designs, builds and operates projects in the sectors of building, infrastructure and industry. As a responsible and committed leader in sustainable construction, Bouygues Construction sees innovation as its primary source of added value: this is "shared innovation" that benefits its customers at the same time as improving its productivity and the working conditions of its 56,980 employees. In 2018, Bouygues Construction generated sales of €12.358 billion.' (Bouygues Construction, 2019).

Following the written complaint, the officers of the Department of Environment conducted onsite inspections and collected samples. The samples showed toxicity and were classified as hazardous.

However, the collection of those samples did not follow the prescribed by law procedures: According to relevant law, the samples should have been divided into three parts, and one part to be given to the accused. This appears to have been a decisive factor in the first instance court decision to acquit Bouygues for all but one of the 12 charges Bouygues was indicted for. However, toxicity was not an element of the charges for the violation of the Solid and Dangerous Waste Management Act of 2002. The first-instance court did not explain why the exclusion of evidence of toxicity, left unfounded the charges for the violation of the Dangerous Waste Management Act of 2002.

The Court of Appeal found that the first instance court, essentially decided without any justification a crucial point in favour of acquitting Bouygues. In its decision, the first instance court failed to analyse the elements of the offences and was criticized by the Supreme Court. Therefore, the Supreme Court decided to allow the appeal against the acquittal of Bouygues. However, it held that, in the light of the intervening period of six years that had elapsed from the relevant facts (2009–2015) and that the counts did not involve felonies, a retrial would not be appropriate in this case.

The case revealed failure by the Department of Environment to collect evidence properly, according to law. The construction of such large environmental impact works such as airports should be under ongoing supervision by the authorities for complying with, and enforcement of high environmental and sustainability standards and slipshod approaches such as those exhibited in this case are inexcusable.

The case also revealed a lack of understanding by the first instance court of such essential concepts as 'waste' and 'pollution', and to analyse the elements of the offences. It also showed a delay in the adjudication of this case: Four years had elapsed from the first instance decision until the appeal. This delay appears to have weighed heavily in the decision of the Supreme Court not to order a retrial. Despite the delay, the Supreme Court decision not to order a retrial gives the wrong messages: that large, international corporations, global players in industries where measures against pollution and protection of the environment must be a primary consideration, can go unpunished.

Last but not least, the failure by the prosecution to service the indictment on three out of four directors resulted in the natural persons to remain unchecked and unpunished. The imposition of monetary fines to global corporations cannot usually have the desired discouragement effect: the fines are internalised as part of doing a large business. The prosecution of natural persons responsible must be a primary consideration, and this is where the weight of relevant authorities should be placed.

5 Concluding Remarks

In conclusion, this chapter has demonstrated how the criminalisation of environmental protection infringements has evolved considerably over the years, at least at EU level, with the hope that the harmonisation of criminal approaches among the Member States would create the necessary incentives for preventing and remedying water pollution. At the same time, by focusing on the situation in Cyprus, this chapter has shown how the enforcement of carefully designed legislation is influenced in practice by various factors, including the capacity for effective prosecution, the coordination of different government departments and the willingness of courts and politicians to uphold the appropriate penalties. In this respect, Cyprus has still a long way to go in enforcing the criminal sanctions that accompany the infringement of a myriad of related EU environmental legislation on water quality and management, starting with the enforcement of the *Environmental Crime Directive* as transposed in the Cypriot legal order. It is high time for those responsible for environmental offences to pay for the damage caused to the environment and affected interests, in line with the well-established polluter-pays principle.

However, for the desired changes vis-à-vis environmental crime in general and water crimes in particular to come about, it is imperative that:

1. personnel in the Water Development Department which is responsible for implementing the water policy of the Ministry of Agriculture, Natural Resources and Environment for the rational development and management of water resources of Cyprus need to be urgently strengthened with specialist young staff (including lawyers) recruited not only for their academic qualifications but, also, for their documented record of having worked as volunteers and so forth for the protection of the environment;
2. the Ministry should set up a group to systematically collect information about water offences and offences in both urban and rural areas in collaboration with the police and inform accordingly the Attorney General who is responsible for all prosecutions in Cyprus. In this context and earning from the local experience with fighting corruption, to establish a hotline and an online facility for citizens to anonymously pass on information about water crime offences and offenders, with the aforementioned group within the Ministry being responsible for evaluating information received and for passing it on to the police.
3. in order to raise awareness of the importance of water resources and the seriousness of water crimes among school-aged children, in collaboration with the Ministry of Education to introduce into the primary and secondary school syllabi material on the importance of protecting water resources from wastage and contamination, for example, and informing the authorities when one becomes aware of such offences and offenders.
4. relevant NGOs and the Ministry to systematically organise awareness seminars both for the judiciary and prosecutors in the different districts as well for members of parliament.

5. in order to raise the general public's awareness about water conservation and the seriousness of water crimes, NGOs and the Ministry to enlist the co-operation of the media for a systematic campaign in the press, on radio, television and in the social media.
6. the government of Cyprus, through its Research Advancement Foundation, to make research funding available on a competitive basis for research into water crimes in Cyprus.

As this chapter shows, water is too important a subject and water crimes too serious to be left to a few low-motivated individuals in different government departments in Cyprus.

References

Alexiades, S. (1981). *The penal protection of the environment as a problem of crime prevention policy (in Greek)*. Athens: A.N. Sakkoulas. Αλεξιάδης, Σ. (1981). Η Ποινική Προστασία του Περιβάλλοντος ως Πρόβλημα Αντεγκληματικής Πολιτικής. Αθήνα: A.N. Σάκκουλας.
Andersen, M. A. (2014). *Environmental criminology: Evolution, theory, and practice*. London and New York: Routledge.
Bouygues Construction. (2019). *A leader in sustainable construction*. Retrieved 31 May, 2019, from https://www.bouygues-construction.com/en/editorial/leader-sustainable
Convention for the Protection of the Mediterranean Sea Against Pollution. (1976). 1102 UNTS 27.
Directive. (1991). Council Directive 91/271/EEC of 21 May 1991 concerning urban waste-water treatment. OJ L 135/40.
Directive. (1998). Council Directive 98/83/EC of 3 November 1998 on the quality of water intended for human consumption [1998] OJ L 330/32.
Directive. (2000). Directive 2000/60/EC of the European Parliament and of the Council of 23 October 2000 establishing a framework for Community action in the field of water policy. OJ L 327/1.
Directive. (2004). Directive 2004/35/CE of the European Parliament and of the Council of 21 April 2004 on environmental liability with regard to the prevention and remedying of environmental damage. OJ L 143/56.
Directive. (2006a). Directive 2006/118/EC of the European Parliament and of the Council of 12 December 2006 on the protection of groundwater against pollution and deterioration. OJ L 182/52.
Directive. (2006b). Directive 2006/7/EC of the European Parliament and of the Council of 15 February 2006 concerning the management of bathing water quality and repealing Directive 76/160/EEC. OJ L 64/37.
Directive. (2008). Directive 2008/99/EC of the European Parliament and of the Council of 19 November 2008 on the protection of the environment through criminal law. OJ L 328/28.
Directive. (2009). Directive 2009/123/EC of the European Parliament and of the Council of 21 October 2009 amending Directive 2005/35/EC on ship-source pollution and on the introduction of penalties for infringements. OJ L 280/52.
Directive. (2010). Directive 2010/75/EU of the European Parliament and of the Council of 24 November 2010 on industrial emissions (integrated pollution prevention and control). OJ L 334/17.
EFFACE Synthesis. (2016). *Synthesis of the Research Project 'European Union Action to Fight Environmental Crime' (EFFACE)*. Retrieved 30 April, 2019, from https://efface.eu/sites/default/files/publications/EFFACE_synthesis-report_final_online.pdf

European Report on Water Crimes. (2019). Retrieved 30 April, 2019, from http://documents.rec.org/publications/WaterCrimesFinalReport.pdf

Farmer, A., Faure, M., & Vagliasindi, G. M. (2017). *Environmental crime in Europe*. Dublin: Hart Publishing.

Fisher, E., Lange, B., & Scotford, E. (2013). *Environmental law: Text, cases & materials*. Oxford: Oxford University Press.

Industrial Emissions Law. (2013). Industrial Emissions (Integrated Pollution Prevention and Control) Law of 2013 (Law 184 (I)/2013).

International Convention for the Prevention of Pollution from Ships (MARPOL). (1973). 1340 UNTS 184.

Kapardis, A. (1983). Use and impact of imprisonment in Cyprus: A quasi-experiment. *Cyprus Law Review, 1*(2), 361–374.

Kapardis, A. (1986). Juvenile delinquency and delinquents in Cyprus. *Cyprus Law Review, 4*, 2371–2379.

Kapardis, A. (2018). Why Cyprus needs green cultural criminology. In A. Kapardis & A. Samara-Krispi (Eds.), *Protecting the environment form human interference in international law* (pp. 125–135). Athens: A. N. Sakkoulas.

Kapardis, M. K. (2001). In A. Kapardis & N. Kourakis (Eds.), *Economic crimes in Cyprus: A multidisciplinary approach*. Athens: A. N. Sakkoulas.

Law. (2012). Law 22(I) 2012 on the protection of the environment through criminal law.

Mitsilegas, V., Fitzmaurice, M., & Fasoli, E. (2016). The relationship between EU criminal law and environmental law. In V. Mitsilegas, M. Bergström, & T. Konstadinides (Eds.), *Research handbook on EU criminal law* (pp. 272–293). Cheltenham, UK: Edward Elgar Publishing.

Pyrgou, M. (2018). Prosecuting environmental law offenders in Cyprus: Principles and case law. In A. Kapardis & A. Samara-Krispi (Eds.), *Protecting the environment form human interference in international law* (pp. 137–151). Athens: A. N. Sakkoulas.

Sutherland, E. (1940). White collar criminality. *American Sociological Review, 5*(1), 1–12.

Theopemptou, C. (2018). The case of 5 crimes against the environment. In A. Kapardis & A. Samara-Krispi (Eds.), *Protecting the environment form human interference in international law* (pp. 153–165). Athens: A. N. Sakkoulas.

UN Convention on the law of the Non-Navigational Uses of International Watercourses. (1997). 36 ILM 700.

UN Convention on the Law of the Sea. (1982). 1833 UNTS. 397.

UNECE Convention on the Protection and Use of Transboundary Watercourses and International Lakes. (1992). 1936 UNTS 269.

United Nations Environment Programme. (2016). The rise of environmental crime: A growing threat to natural resources. *Peace, Development and Security*. Retrieved 5 June, 2019, from http://wedocs.unep.org/handle/20.500.11822/7662

Vagliasindi, G. M. (2017). The EU environmental crime directive. In A. Farmer, M. Faure, & G. M. Vagliasindi (Eds.), *Environmental crime in Europe*. Hart Bloomsbury Publishing.

West, L, 2006). *World Water day: A billion people worldwide lack safe drinking water*. Retrieved 5 June, 2019, from http://www.about.com

World Economic Forum. (2015). *Global Risks*.10th ed. Retrieved 5 June, 2019, from http://www.weforum.org/docs/WEF_Global_Risks_2015_Report15.pdf

Chapter 9
Water Protection in Slovenia: Constitutional and Administrative Law Perspectives

Iztok Rakar, Bojan Tičar, and Tina Sever

1 Introduction

Slovenia is a country abundant in water resources, though they are not uniformly distributed across the country (OECD, 2015c). Water is not only an important component of the environment, necessitating policy and measures to ensure its quality and renewal, but also a key factor in social and economic development, also requiring adequate policy for its sustainable use (Akhmouch & Correia, 2016; Briscoe, 2011; UNESCO, 2015). It is, therefore, unsurprising that legal regulation for water protection and its governance is complex (Brilly, 2000; Schophaus, 2019). This chapter presents the constitutional and administrative law perspectives on water protection in Slovenia, and pursuant to its complexity, we focus on selected issues. In terms of constitutional law, we focus on water as a human right in domestic and international law, from the administrative perspective, we focus on water management systems and water protection. Water protection is conceptualised as an *ex ante* and *ex post* activity subject to public and private law to ensure water quality, quantity and accessibility.

The chapter is accordingly structured, as follows: firstly, we provide an overview of the international and constitutional foundations of water protection, with special focus on the right to drinking water; next, we examine the administrative law framework for water protection and provide an overview of its form; we conclude with a brief assessment of water protection legal regulation and practice in Slovenia.

I. Rakar · T. Sever
Faculty of Public Administration, University of Ljubljana, Ljubljana, Slovenia
e-mail: iztok.rakar@fu.uni-lj.si; tina.sever@fu.uni-lj.si

B. Tičar (✉)
Faculty of Criminal Justice and Security, University of Maribor, Ljubljana, Slovenia
e-mail: bojan.ticar@fvv.uni-mb.si

K. Eman et al. (eds.), *Water, Governance, and Crime Issues*,
https://doi.org/10.1007/978-3-030-44798-4_9

2 The International and Constitutional Foundations of Water Protection

2.1 Introduction

Water is a natural resource essential for the life of all beings and ecosystems functioning; it is as part of the natural environment and legally protected by national, supranational and international legislation: constitutions, general environmental protection law, sector-specific law, subordinate legislation, EU legislation and international treaty (*infra,* Fig. 9.1).

Regulation, in the main, is anthropocentric, protecting the right of humans to water; the state is obliged to protect this right by enabling access to water, protecting its quality, and the like. However, as mentioned above, regulation is not merely a national issue; it has an international dimension, including an environmental aspect (Dupuy & Viñuales, 2016).

Consequently, the right to water was gradually introduced in different international documents. Although the right to drinking water was not defined as a *human* right, such nature could be derived from other general human rights, such as the right to life, dignity, and health (Dupuy & Viñuales, 2016; Sancin, Pucelj Vidovič, Kovič Dine, & Vrbica, 2015).

2.2 International Treaties and Documents on the Right to Drinking Water

One of the first international documents including the right to drinking water was a Convention on the Elimination of All Forms of Discrimination against Women (UN General Assembly, 1979), stating that states parties should take all appropriate

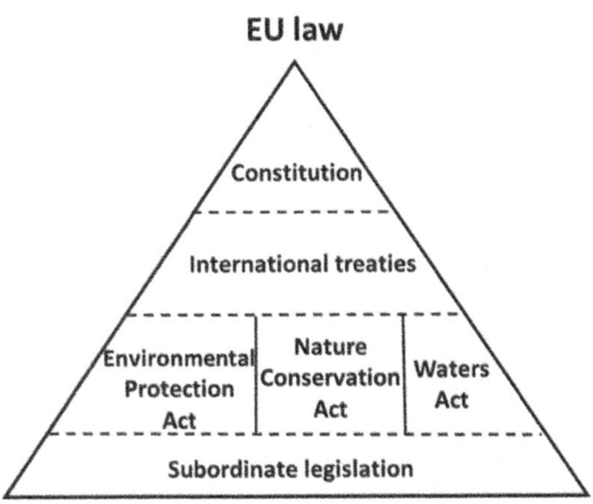

Fig. 9.1 Water protection legislation in Slovenia (Source: authors based on legislation)

measures to eliminate discrimination against women in rural areas and, in particular, ensure adequate living conditions, particularly in relation to housing, sanitation, electricity and water supply, transport and communication.

Moreover, in January 1992, the International Conference on Water and Sustainable Development (Dublin Conference) accepted Principle 4, stating "… it is vital to recognise first the basic right of all human beings to have access to clean water and sanitation at an affordable price". Furthermore, in June 1992, the United Nations Conference on Environment and Development (Rio Summit) in Chap. 18 of Agenda 21 confirmed the Resolution of the Mar del Plata Water Conference (United Nations Water Conference 1977), according to which, all people have the right to have access to drinking water ("the commonly agreed premise").

Additionally, the UN General Assembly adopted Resolution A/Res/54/175, "The Right to Development" (1999), stating that "in the full realization of the right to development, inter alia: (a) The rights to food and clean water are fundamental human rights and their promotion constitutes a moral imperative both for national Governments and for the international community" etc.

Undoubtedly, the most important binding international treaty, from which the human right to drinking water can be derived, is the International Covenant on Economic, Social and Cultural Rights (19 December 1966) (ICESR 1966), which encompasses rights unattainable without access to clean drinking water, that is, the right of everyone to an adequate standard of living for himself and his family, including adequate food, clothing and housing (Article 11) and right to health (Article 12). Moreover, General Comment No. 15 to the ICESR (2002)[1] provides more detail on said rights in articles 11 and 12 of the ICESR. Even though this is merely soft law, its explanation of ICESR, a mandatory law, is an important guideline to states. As stated in comment 15, "The human right to water is indispensable for leading a life in human dignity. It is a prerequisite for the realisation of other human rights" (Article I.1). This clearly shows the interconnection of general human rights and access to water.

Finally, an important turning point in terms of recognition of the right to drinking water as a human right was UN General Assembly Resolution 64/292 on "The Human Right to Water and Sanitation" (UN General Assembly, 2010), which explicitly "recognises the right to safe and clean drinking water and sanitation as a human right that is essential for the full enjoyment of life and all human rights." Although it is not binding, it clearly evidences the importance of the subject matter for states globally.

[1] General Comment No. 15: The Right to Water (Arts. 11 and 12 of the Covenant) Adopted at the Twenty-ninth Session of the Committee on Economic, Social and Cultural Rights, on 20 January 2003 (Contained in Document E/C.12/2002/11), http://www.refworld.org/pdfid/4538838d11.pdf

In signing and ratifying international treaties, states are obliged to respect them; moreover, soft law, such as different resolutions, importantly influences national regulation, providing guidelines on means and will, e.g. water regulation.

The European Convention on Human Rights (ECHR 1950), as well as recommendations, resolutions and conventions of the Council of Europe, is relevant for European states, Recommendation Rec (2001)14 of the Committee of Ministers to member states on the European Charter on Water Resources and Resolution 1693 (2009) by the Parliamentary Assembly of the Council of Europe being two such examples. The ECHR is one of the more important documents for protecting human rights in Europe; the right to water can be derived from Article 2's protection of the right to life. However, the European Court of Human Rights (ECtHR), which interprets the convention, has mostly dealt with water issues in relation to the prohibition of torture (Article 3 of the ECHR)[2] and in relation to private property groundwater pollution pursuant to infringement of the right to fair trial (Article 6 of the ECHR).[3] In terms of the right to a healthy environment, to some extent also related to water protection, the ECtHR, for example, positions environment within the right to respect for private and family life (Article 8 of the ECHR).[4]

Finally, EU member states are bound by EU law, which posits water regulation within environmental and consumer law (Thielbörger, 2014). However, water resource ownership and means to provide water supply fall under individual member state regulation (Pekolj, 2014). Similarly, at the EU level, the right to drinking water can be derived from the Charter of Fundamental Rights of the European Union (Official Journal C 364, 18 December 2000), protecting rights, such as human dignity and the right to life. The following directives, furthermore, are relevant in terms of water regulation: "Council Directive 98/83/EC of 3 November 1998 on the quality of water intended for human consumption (Official Journal L 330, 5 December 1998)", "Directive 2000/60/EC of the European Parliament and of the Council of 23 October 2000 establishing a framework for Community action in the field of water policy (Official Journal L 327, 22 December 2000)" supplemented by "Directive 2006/118/EC of the European Parliament and of the Council of 12 December 2006 on the protection of groundwater against pollution and deterioration (Official Journal L 372, 17 December 2006)" and "Directive 2008/105/EC of the European Parliament and of the Council of 16 December 2008 on environmental quality standards in the field of water policy (Official Journal L 348, 24 December 2008)". The provision of drinking water was excluded from Directive 2014/23/EU of the European Parliament and of the Council of 26 February 2014 on the award of concession contracts (Official Journal L 94, 28 March 2014) pursuant due to strong civil opposition (see Article 40).

---

[2] Case of Elci and Others v. Turkey (Applications nos. 23,145/93 and 25,091/94), 13 November 2003. Case of Ostrovar v. Moldova (Application no. 23145/93), 13 September 2005.

[3] Case Zander v. Sweden (Application no. 14282/88), 25 November 1993.

[4] E.g. Case Brânduşe v. Romania (Application no. 6586/03), 7 April 2009.

2.3 *Slovenia's Constitutional Foundation for Water Protection*

Slovenia is water resource rich,[5] but this should not be taken for granted; on the contrary, water resources should be adequately protected for future generations in accordance with the principle of sustainability through the application of multilevel water governance to enable public participation and action (see *infra,* Chap. 3.4). An important means for such protection is appropriate regulation, followed by efficient implementation.

In accordance with Article 8 of the Constitution, Slovenia's signing of international treaties obliges it to respect its rights and obligations,[6] one being the right to drinking water. To explicitly define and protect it at a constitutional level, the right to drinking water was incorporated in the Constitution by adding article 70 a. ("2016 amendment", see *infra* Chap. 2.4). However, legal protection of water resources and the right to drinking water could be derived from constitutional provisions enabled prior to 2016's amendment and sector-specific legislation.

Article 72 of the Constitution protects the right to a healthy environment, essential for the implementation of human rights, such as the rights to life, personal integrity, family life and health (UNESCO, 2012). Despite being in Chap. III of the Constitution, which only includes legal standards and program norms, it is regarded as a third generation solidarity and collective right with the status of constitutionally protected human right (Šturm, 2011). Therefore, taking into account Articles 15 (exercise and limitation of rights) and 8 (direct application of ratified and published international treaties), in accordance with Articles 17 (right to life), 21 (protection of dignity) and 72 (right to healthy living environment), the *right to drinking water* was already protected before 2016 (Sancin et al., 2015).

Articles 72 and 70 (national assets and natural resources) are relevant for the *management of water resources* (Sancin et al., 2015); namely, in accordance with the right to healthy environment, it is the duty of the state to legislatively define the conditions and means to perform economic activity. This was facilitated by adoption of the Environmental Protection Act (2004 and amendments), Services of General Economic Interest Act (1993 and amendments), Nature Conservation Act (1999 and amendments), Waters Act (2002 and amendments), Spatial Planning Act (2017) and Animal Protection Act (1999 and amendments) amongst others. Drinking water supply was defined as an obligatory local public service, which could be performed by public utilities, companies and service agencies, or through private entity concession where profit is subordinate to the satisfaction of public need.[7]

Furthermore, Article 70 defines possibilities for the special use of public goods and exploitation of natural resources as defined by law. In addition, Article 71

[5] Water covers about 272 km^2 of Slovenia's territory, while geologic units, which may store and transmit groundwater, contain about 50 m^3/s of dynamic supply (OECD, 2015c).

[6] E.g. Slovenia is obliged by the ICESR since 1 July 1992 (Official Gazette of the Republic of Slovenia No. 35/92 - MP, No. 9/92).

[7] More on drinking water supply in Slovenia in Pevcin and Rakar (2018).

defines land protection in accordance with conditions set by law. When establishing conditions, legislators should respect Article 67 of the Constitution on property and regulate land use to ensure its economic, social, and *environmental* function in accordance with the public interest. However, it should not be defined in such a way as to diminish the right to property (Šturm, 2011).[8]

These provisions are also important for water land management and other rights related to water, as defined by sector specific legislation, such as the Waters Act (for detailed analysis on sectoral legislation see Chap. 3). As derives from Article 33 the right to private property is a human right. The Constitutional Court of the Republic of Slovenia, besides protecting property, protects individuals, enabling freedom of performance in terms of property (Decision No. Up-156/98, 11 February 1999). The Constitutional Court has stated that interruption of water supply to apartments constitutes interference with the right to private property (Decision No. Up-156/98, 11 February 1999), and did not focus on right of access to water (Knez, 2015a).

Finally, Article 73 explicitly defines the duty of the state and local authorities to promote the preservation of natural heritage. This provision implements the duty of the state to protect human rights on its territory and appropriately manage natural resources, as defined by Article 5 of the Constitution.[9] The right to protect natural heritage is a positive right, placing the burden of adequate activity on the state and directly interrelated with the human right to a healthy environment (Letnar Černič & Čebulj, in Šturm, 2011). In accordance with the Article 15 of the Constitution, human rights are exercised directly based on the constitution, but legislation can prescribe manner of rights implementation, e.g. the Nature Conservation Act (NCA),[10] which defines nature preservation, including water falls, lakes, wells, and rivers (see Article 4 of NCA).

2.4 *The Right to Drinking Water Pursuant to 2016's Amendment to the Slovene Constitution*

Despite the arguments that water resources were already efficiently protected by the Constitution and sectoral legislation; ultimately, Article 70's provisions, amongst the most important, together with Article 72, for water resource protection, were assessed as problematic, being too "loose" and not defining natural resource composition or setting conditions for the use of said resources, such as water. This is left to the discretion of legislators, who can be broad in their approach when addressing these questions. Moreover, Article 70 states that natural resources can also be exploited by foreign persons, giving rise to concern about neoliberal water resource

[8] Cf. Constitutional Court Decision No. U-I-107/96, 5 December 1996.

[9] See e.g. Constitutional Court Order No. U-I-226/04–36, 1 December 2005.

[10] Official Gazette of the Republic of Slovenia No. 56/99 with amendments.

privatization, and concomitantly encouraging civil society to take a stand;[11] consequently, the Slovene Constitution was amended by the acceptance of the Constitutional Act adding Article 70.a to Chapter III of the Constitution, which explicitly defines and protects the right to drinking water (Official Gazette of the Republic of Slovenia No. 75/16, 30 November 2016).

Chapter III of the Constitution regulates economic and social relationships, including program norms and legal standards, usually requiring legislative and case law content definition; Article 70.a protects the right to drinking water as a human right (based on Article 5 of Constitution) and defines positive and negative state obligation; the state must not interfere rights but must create exercise conditions (Letnar Černič, in Avbelj, 2019). Article 70.a's defines the *right of every individual to drinking water* (1) and *water resources as a public good,* and confers *state level management* (2); it, moreover, states that water resources are *primarily and sustainably for the supply of drinking water* to the population, and supply to households and are in these respect *not a tradable good* (3); finally, it declares that *the state ensures drinking water supplies directly and non-profitably through local self-government communities* (4).

With this explicit incorporation of the right to drinking water into its constitution, Slovenia became one of the few countries, such as South Africa, Uruguay, India and Slovakia, to protect this right at the constitutional level. Of course, this is only the first step, legislation harmonization must follow; moreover, the amendment stirred up questions that need to be answered by sectoral legislation.

Article 70.a's first paragraph confers the right to drinking water to everyone on Slovene territory (Letnar Černič, in Avbelj, 2019), and the state is obliged to enable this right; however, it does not oblige the state to provide water supply in the areas where self-sufficiency is the rational choice or if real estate cannot be connected to water infrastructure (Report on the Preparation of Constitutional Act on Supplementing Chapter III of the Slovene Constitution, 2016).

Water resources for supply to the population are dealt with in paragraph two, whereby resources, be they natural or artificial, are a public good under state management,[12] not ownership. The state is additionally responsible for protecting water resources, including monitoring water status, preventing and reducing pollution, encouraging sustainable water use, and protecting water environment (Report on the Preparation of Constitutional Act on Supplementing Chapter III of the Slovene Constitution, 2016).[13]

Exercise of the right to drinking water is provided for in paragraph three: "water resources, serving as drinking water supply for the population and supply of the households are *in this part* not tradable goods"; however, such definition is problematic pursuant to possible *argumentum a contrario*, which raises the question of

[11] Civil Initiative for Slovenia and Freedom.

[12] See assenting opinion from judge Accetto to Constitutional Court Decision No. U-I-164/14–31, 11 December 2017.

[13] See assenting opinion from judge Accetto to Constitutional Court Order No. U-I-107/16–10, 1 March 2018 on protection of environment, also by adding right to drinking water in the Constitution.

water resources not intended for supply. Do they automatically become tradable goods? Moreover, what is a "*water resource*"? Gantar (2016) defines water resources as surface and below the ground water supply sources. Water resources, as such, should never become tradable goods, even if they are not meant for drinking water supply unless deemed tradable by permit[14] or concession (Gantar, 2016).

In accordance with the constitutional principle of free economic initiative (Article 74 of Constitution), 2016's amendment still enables the exploitation of water resources for special use, such as beverage production; however, as mentioned above, the state has to confer the necessary permits and concessions (Report on the Preparation of Constitutional Act on Supplementing Chapter III of the Slovene Constitution, 2016). Water permits and concessions are still regulated by current Waters Act (Official Gazette 67/02 and amendments), which stipulates that drinking water supply takes precedence over the use of water for other purposes (Article 108); should there be water regime change leading to shortage, those with permit and concession are obliged to adjust their use accordingly (Article 112).

Future concession contracts for economic exploitation should contain provisions on the impact of the preferential and sustainable supply of drinking and household water to the population (Report on the Preparation of Constitutional Act on Supplementing Chapter III of the Slovene Constitution, 2016) in accordance with 2016's amendment.

Article 70.a's final paragraph lacks clarity in its definition of state and local competence with regard to supplying drinking water: it is not clear how the state will directly supply drinking water through local self-government communities.[15] The Environmental Protection Act and Local Self-Government Act deemed local self-government communities competent for water supply, but according to 2016's amendment, this is no longer the case. The deadline for legislative and constitutional alignment of May 2018 has passed and alignment has not taken place (see more below). According to Article 70.a, the state manages water resources, and is responsible for their protection and for non-profit water supply as obligatory public utility services (Report on the Preparation of Constitutional Act on Supplementing Chapter III of the Slovene Constitution 2016; Pličanič, 2017); local communities supply water when possible; if not, the state ensures supply (Report on the Preparation of Constitutional Act on Supplementing Chapter III of the Slovene Constitution 2016). The constitution defines non-profit drinking water supply to the population to eliminate activity concession, and as such, water supply is not defined as a market-led commercial activity: public service must be publicly provided, price set to enable public service provision, investment, and the like, any excess funds used solely for water supply purpose (Report on the Preparation of Constitutional Act on Supplementing Chapter III of the Slovene Constitution 2016; Pevcin & Rakar, 2018). However, despite the fact that public service is provided publicly, this

[14] Altogether 38,729 water permits were issued (includes water permit for all types of water use as defined by Article 125 of Waters Act).

[15] Constitutional Court Order No. U-I-121/19–7, 13 June 2019.

does not ensure non-profit performance (Senčur in Report on the Preparation of Constitutional Act on Supplementing Chapter III of the Slovene Constitution 2016; Knez, 2015b).

At least 15 laws need to be aligned in accordance with 2016's amendment (Kralj, 2017). The government appointed an inter-ministerial working group to facilitate alignment by the end of May 2018, and it prepared an overview of laws needing amendment, that is the Environmental Protection Act, Services of General Economic Interest Act, and Local Self-Government Act (Ministry of Environment and Spatial Planning, 2018). To do so, correct interpretation of paragraph 4 of Article 70.a is needed to define the relationship between state and local community with regard to drinking water supply. Consequently, the government appointed a new inter-ministerial working group in December 2018 to prepare guidelines on legislative amendments by the end of December 2019; regulation in force since before the adoption of the Constitutional Act (Official Gazette of the Republic of Slovenia No. 75/16, 30 November 2016) apply until new legislation is adopted.

Pursuant to our analysis above, the following chapters will analyse sector-specific water protection legislation and water supply's constitutional provision through administrative law, administrative law being a democratic tool for providing constitutional safeguards against power abuse (Sever, Rakar, & Kovač, 2014); moreover, administrative substantive law is implemented in accordance with the public interest.

3 Administrative Law Framework for Water Protection

3.1 Introduction

Legal norms stipulate standards for the behaviour of individuals and institutions (*ex ante* dimension of law), against which, actual behaviour is measured (*ex post* dimension of law) (van Kralingen 1997; Pavčnik, 2015). Slovenia's Environmental Protection Act of 2004 (EPA) and Nature Conservation Act of 1999 (NCA) being *lex generalis,*[16] its Waters Act of 2002 (WA) being *lex specialis* with respect to water protection legislation (Fig. 9.1).[17] This legislation is the foundation of subordinate legislation issued by the government and municipalities.[18]

[16]According to Berginc, Kremesec-Jevšenak, and Vidic (2007) NCA is *lex specialis* in relation to EPA.

[17]Basic policy documents of water management are resolutions on national environmental action plans. The last resolution covered the period 2005–2012, whilst the one for the period until 2030 is still in the preparation phase.

[18]Municipalities' e. g. issue decrees on the protection of sources of drinking water and on traffic regimes on lakes.

EPA, NCA and WA contain substantive, organisational and procedural norms. The latter are *lex specialis* in relation to the General Administrative Procedure Act of 1999.

EPA defines water as part of the environment (Article 3 of EPA) and regulates the following water related issues: strategic environmental assessment (Article 40), environmental impact assessment (Article 51), environmental protection approval (Article 57) environmental protection permit (Articles 70 and 78), monitoring (Article 96), environmental pollution and damages (Articles 110.a and 110.b), and environmental protection supervision service (Article 159).

NCA establishes a system for protecting natural assets, including lakes and rivers, to preserve natural asset and process characteristics, and to restore natural assets (Article 4).

Water protection is part of water management (Fig. 9.2) according to WA; however, from a broader perspective, these provisions only represent *ex ante* water protection *sensu stricto*, whilst *ex ante* water protection *sensu lato* pertains to: (1) legal regulation of water resource use; and (2) technical water regulation (Pličanič, 2017). This follows, *inter alia*, from the basic principles of water management, most directly from the long-term protection of high quality water and its proper use, and environmental protection (Article 3). Accordingly, technical regulation of water pertains *inter alia* to: (1) water quantity preservation; (2) water and waterside land maintenance; and (3) maintenance of water regime hydro-morphological condition (Article 80).

We may therefore speak of *ex ante* water quality and quantity protection (cf. Woodhouse & Muller, 2017). *Ex post* water protection also pertains to supervision, obligatory and voluntary (*infra*, Chap. 4).

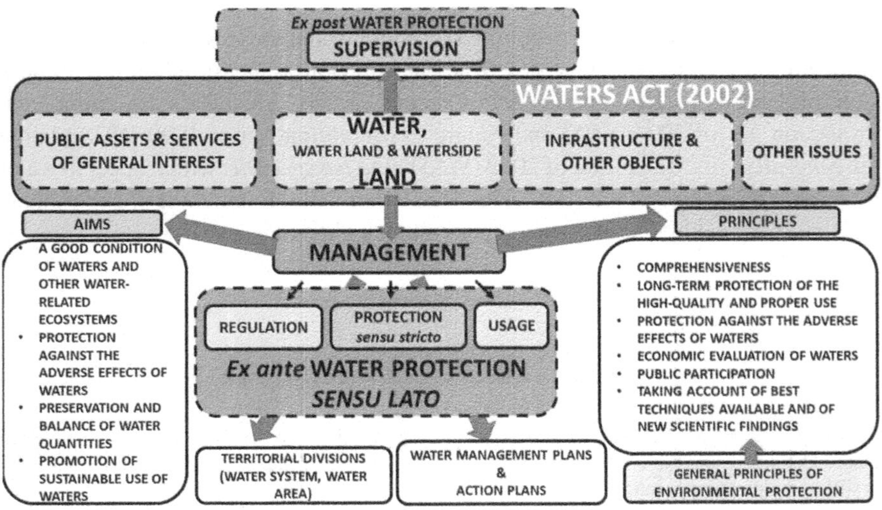

Fig. 9.2 Water protection according to the Waters Act of 2002 (Source: authors based on WA)

Water protection *sensu lato* is based on: (1) the territorial division of Slovenia in terms of water systems and areas (Articles 52 and 53); (2) the national plan adopted by parliament (Article 54), and water management and action plans adopted by the government (Articles 55–59).[19]

Spatial planning acts[20] should also be aligned with water management plans, such as spatial plans for water protection being acknowledged and ministry set guidelines and approval of spatial planning if guidelines are followed (Article 61).[21]

3.2 *Institutional Setting*

Comparatively, answers to the basic questions related to water governance[22] are not uniform (Ozerol et al., 2018); Woodhouse and Muller (2017) find water governance highly contextual, influenced by physical factors, and economic and social development, as well as political and cultural norms that have often evolved over long periods of time; therefore, no common architecture can be demonstrated (Woodhouse & Muller, 2017).[23]

In Slovenia, competence and tasks pertaining to water protection *sensu lato* are distributed amongst several bodies and organisations; accordingly, we can speak of public-private institutional mix (Fig. 9.3; cf. OECD, 2015c);[24] institutional arrangement of water protection *sensu stricto* includes public and private legal organisations, too. The former pertains to central government agencies and providers of services of general interest, the later to individual natural persons with public authority (*infra*, Chap. 4.3).

Institutional complexity is a challenge for efficient and effective water protection. OECD Principles on Water Governance (OECD, 2015a) state that coordinated cross-sector water resource protection management is key for policy coherence and, ultimately, effective water management, that is protection *sensu lato*.

---

[19] Available at https://www.gov.si/teme/nacrt-upravljanja-voda-na-vodnih-obmocjih/ (accessed 3 Sept 2019).

[20] Spatial plans are *sui generis* legal form (Köck, 2012).

[21] Critically in terms of the normative perspective and its implementation in practice (Pličanič, 2017).

[22] I.e., who should participate in decision-making; at what geographical and political level should governance institutions operate; and what is the appropriate role of market and non-market criteria in relation to water allocation.

[23] Nevertheless, some generic governance system features emerge (Woodhouse & Muller, 2017). Similarly, according to Mechlem (2016), institutional set-ups for managing aquifers and administering water rights vary greatly globally. In most countries, a large number of multilevel water resource administration institutions exist (Nanni, 2007; Akhmouch & Correia, 2016).

[24] The former pertains, for example, to ministry, agency and public institute, the later, for example, to private companies.

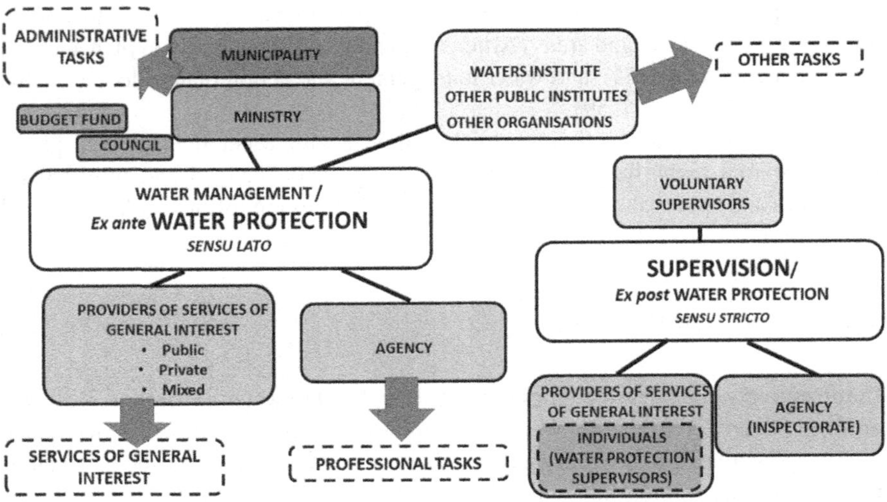

Fig. 9.3 Water protection institutional setting (Source: authors based on WA)

3.3 Water Protection Sensu Stricto

Water protection *sensu stricto* consists of: (1) prohibitions and restrictions; (2) water body classification; and (3) protected areas (WA, part II, Chap. 3).

Water body classification is based on legislative environmental parameters and determined by ministers' rules.[25] Prohibitions and restrictions pertain to: (1) wastewater discharge (Article 64); (2) fertilisation and plant protection agents (Article 65); (3) navigation practices (Article 66), (4); washing (Article 67); (5) depositing or disposal of substances and objects (Article 68), (6) hazardous substances (Article 69); (7) notification and action (Article 70); (8) water flow (Article 71); (9) alluvium (Article 72); and (10) estuarial waters (Article 73). Rules on protected areas pertain: to (1) protected water areas (Sect. 3.3.1); (2) bathing water areas (Sect. 3.3.2); (3) designation (Sect. 3.3.3); (4) compensation (Sect. 3.3.4); and (5) protected surface water areas (Sect. 3.3.5).

These provisions reveal WA's complexity of objects, subjects and legal techniques for its water protection regime (Table 9.1; Fig. 9.4).[26]

Water management is, in principle, a central government competence, administrative tasks performed by local branches of the central government agency responsible for water (*supra*, Fig. 9.3). Only in WA specified tasks are municipal competence, e.g. adopting detailed plans of water management (WA 2002: Articles 4 and 90; *supra*, Table 9.1).[27]

[25] See, for example, Rules on determining and classification for water bodies on surface water, Official Gazette of the Republic of Slovenia, No. 63/2005 and amendments.

[26] For detail elaboration of regulatory techniques see Pečarič (2016).

[27] For detailed overview of municipal competence see Lex localis (2018).

Table 9.1 Water protection *sensu stricto* (Source: authors based on WA)

Subject matter	Regulator	Legal norm addressee	Legal technique/act	Water Act article/section
Classification	Parliament, Government, Minister	Minister	Rules (subordinate legislation)	62, 63
Wastewater discharge, fertilisation and plant protection agents	Parliament, Government, Minister	General public	Law and subordinate legislation	64, 65
Navigation practices, exemptions	Parliament	General public/ inland water users	Law	66
	Government		Subordinate legislation	
Washing	Parliament	General public	Law	67
Depositing and disposal of substances and objects				68
Hazardous substances: production, use, storage, handling and transport, and facility and plant construction	Minister	General public, businesses	Subordinate legislation	69
	Parliament		Law	
Notification and action	Parliament	General public	Law	70
Water flow	Parliament	Holders of special rights of water usage, holders of water rights	Law	71
Alluvium	Parliament	General public, providers of services of general interest, holders of special water usage rights	Law	72
Estuarial waters	Parliament	General public	Law	73
Protected water areas	Parliament	Government	Regulations	Sect. 3.3.1
		Minister	Rules	
		Municipality (subsidiary government)—building facilities for water supply	Law, spatial plan (local and subsidiary central)	
		General public, businesses, land owners and beneficiaries	Law	
	Central government—deconcentrated agencies	Land owners	Decision	

(continued)

Table 9.1 (continued)

Subject matter		Regulator	Legal norm addressee	Legal technique/act	Water Act article/ section
Bathing water areas	Quality management	Government	Public and private bodies	Regulations	Sect. 3.3.2
		Minister		Rules	
	Activity limitation	Parliament	Land owners and beneficiaries	Law	
		Central government—deconcentrated agencies		Decision	
Marking		Parliament	Provider of services of general interest, holders of water rights, municipality	Law	Sect. 3.3.3 Article 78
			Minister	Rules (subordinate legislation)	
Compensation		Parliament	Water land managers (ministry), landowners, providers of drinking water supply	Contract	Sect. 3.3.4 Article 79
		Government	Holders of agricultural activity	Subordinate legislation	
Protected surface water area	Determination of areas	Parliament	Government	Subordinate legislation	Sect. 3.3.5 Article 79.b
	Activities restriction and prohibition, obligation to performing or facilitate measures	Parliament, Government, administrative agencies	Landowners, land possessors	Law, subordinate legislation, decision	
	Expropriation and limitation of property rights	Central government—deconcentrated agencies		Decision	
	Risk analysis	Minister		Subordinate legislation	

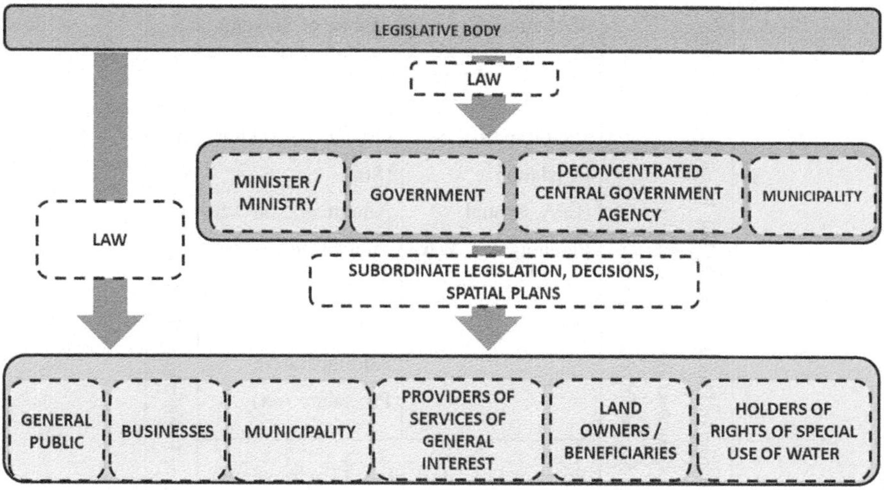

Fig. 9.4 Bodies and water protection legislation (Source: authors based on WA)

3.4 Public Participation

Public participation is a fundamental principle of environment and water protection, and good administration and good (local) governance concepts (cf. Council of Europe, 2008; Godden & Ison, 2019). According to the literature, public participation serves many goals (Rakar, 2011).[28] The OECD (2015a), for example, views participation as an instrument for achieving more effective, efficient and inclusive water policy design and implementation. Nevertheless, water legislation and governance complexity means there are many challenges to be met (Woodhouse & Muller, 2017).

Public participation in Slovenia has been embedded in administrative reform and modernisation since the early 2000s (Virant & Rakar, 2017). Slovene legislation and soft law Slovenia follows international open and transparent governance trends (Scheme 9.1); however, an implementation gap exists. According to OECD findings published in 2015, Slovenia had room for improvement, its Regulatory Policy receiving an average grade in terms of public participation in law making and *ex ante* regulatory impact assessment, its grade for *ex post* regulatory impact assessment was lower (OECD, 2015b). The OECD states that Slovenia's government has made great strides in improving its regulatory policy in its latest report on Regulatory Policy; however, challenge remains in terms of ensuring that line ministries implement regulatory policy tools, such as regulatory impact assessment and stakeholder engagement, effectively (OECD, 2018; Kovač, 2017; Rakar & Kovač, 2019).

[28] Public participation may be conceptualised as an *ex ante* mechanism for water protection. This type of protection could be essential, since *ex post* mechanisms do not always deliver results in achieving the aims of (legal) regulation pursuant to the many factors influencing efficiency and effectiveness (see Rakar, Tičar, & Sever, 2017).

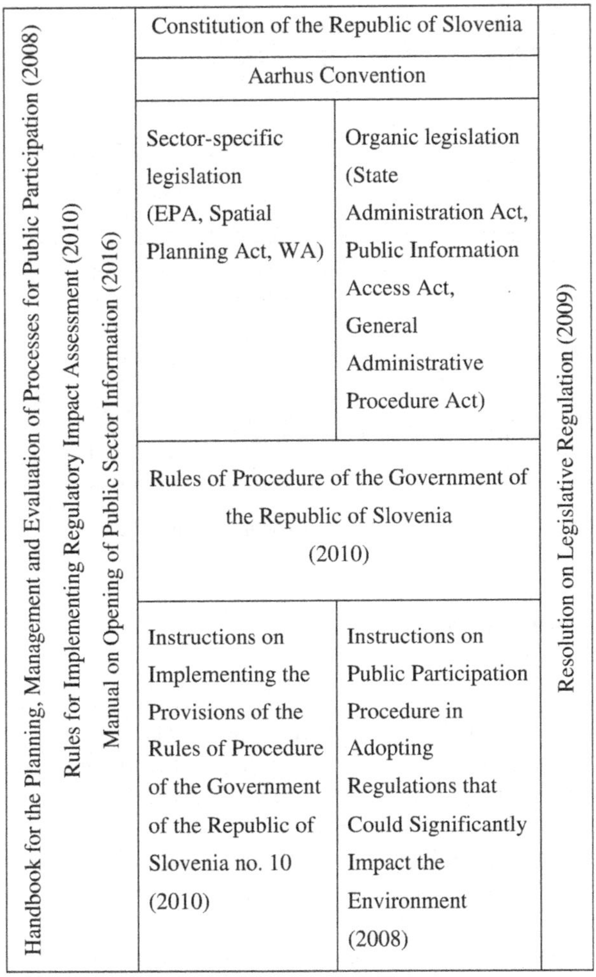

Scheme 9.1 Public participation in law making (Source: Authors based on legislation)

Slovenia's cooperative arrangement's failure to establish water conferences needs mentioning; according to WA, water conferences should have been established to enable local communities, holders of water rights and non-governmental organisations to influence water management (Article 163), being a basic governance mechanism (Biermann, 2007). However, this organisational form is not evidenced in practice, despite financing being made available in the state budget. According to the Government, the provisions on public participation in legislative and decision-making procedures did not accord with democratic standards and EU law and were inadequate duplication of provisions on public participation from Article 58 of WA (see *infra*; Government of Republic of Slovenia 2012). Therefore, 2012's WA amendment abolished this form of public participation and governance.

Public participation, according to WA, pertains to water management planning procedure (Article 58), which is in line with the clear trend of increasing

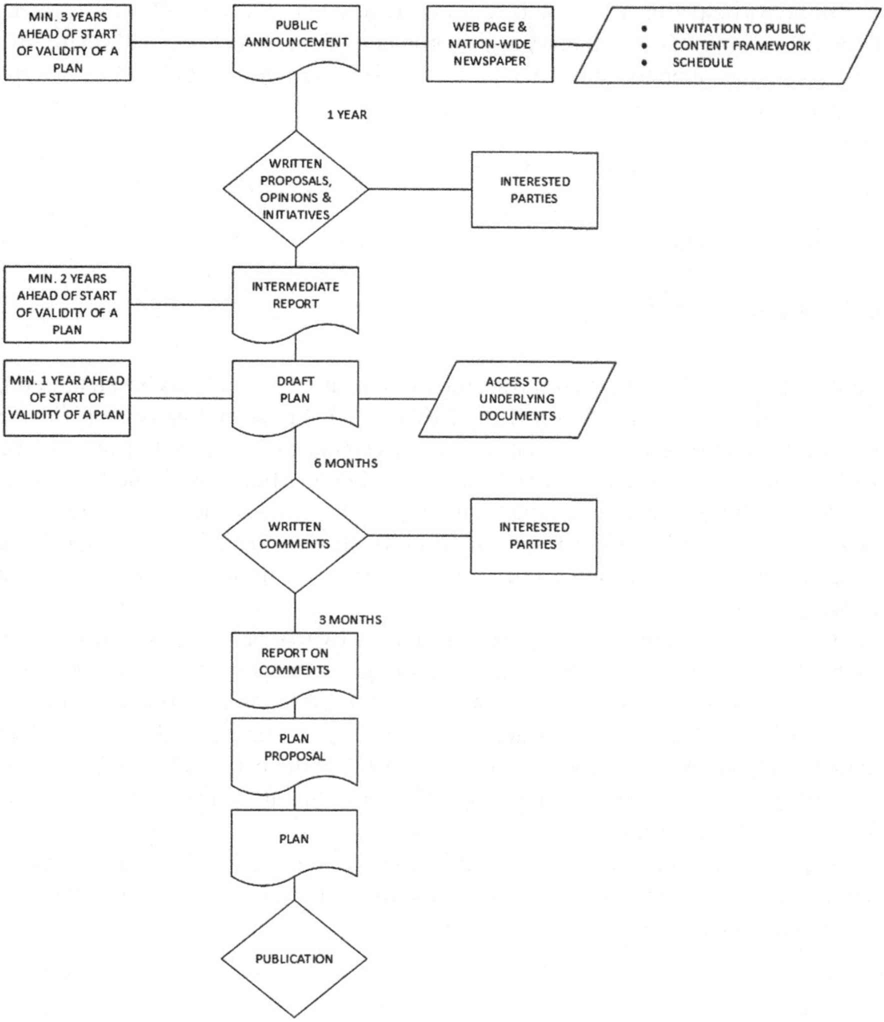

Scheme 9.2 Public participation in water management planning (Source: authors based on legislation)

involvement of stakeholders in plan preparation (Mechlem, 2016).[29] In line with Scheme 9.2, the public is involved at the early stages of decision-making processes and has access to materials used as foundation for draft plan preparation.[30]

[29] Under Art. 14 of the EU Water Framework Directive, Member States shall encourage the active involvement of all interested parties in the implementation of the EU Water Framework Directive and in particular in the production, review, and updating of river basin management plans, which are mandatory for all basins under the directive (Mechlem, 2016).

[30] For the phases of the law making process in Slovenia from draft law to adoption by the National Assembly, see OECD (2018) and Virant and Rakar (2018).

Public participation also broadly pertains to a special form of WA water protection supervision that is voluntary water protection supervisors; we will, therefore, next focus our attention on water protection supervision as regulated by sector-specific law.

4 Supervision

4.1 Introduction

Public interest, as defined in the WA and implementing regulations based thereon, is a key element of supervision (Kovač Kovač, 2016a, b), and protected if those on whom water protection legislation imposes rights and duties act in accordance with norms. The primary responsibility to act in compliance with the law therefore lies with liable persons, supervision representing a secondary responsibility (Jerovšek & Kovač, 2008). From an administrative point of view, supervision ensures feedback and consequent basis for new policy cycles (Tičar & Rakar, 2011).

In accordance with the WA, we distinguish two basic forms of supervision (Table 9.2): obligatory and voluntary. The obligatory is performed under the auspices of: (1) central government agencies by inspectors; and (2) providers of services of general interest by water protection supervisors as holders of public authority (Rakar & Tičar, 2017; Constitution 1991: Article 121; WA 2002: Articles 177–180). Voluntary supervision is performed by voluntary water protection supervisors (WA 2002: Article 180.a).

Additional supervision is performed by the financial authorities: supervision of payment for water rights and water compensation. In these cases, financial and tax legislation applies (WA 2002: Article 180.b).

Table 9.2 Water protection supervision (Source: authors based on WA)

Type	Obligatory		Voluntary
1. Organisation	Central government (agency)	Providers of services of general interest	
2. Title	Inspector	Water protection supervisor	Voluntary water protection supervisor
3. Status	Civil servant	Holder of public authority	
4. Legal basis	Employment under Civil Servants Act (2002)	Administrative decision, and labour or other contract	Administrative decision
5. Competence	Full	Limited	Limited

4.2 Inspection

Inspection is the basic form of water protection legislation supervision, and performed by civil servants within inspectorates.[31] Inspectorates are central government agencies organised within ministries, and territorially organised through regional branches and local offices. Within the framework of inspection, the statutory power to supervise is primarily granted to inspectors competent for water within the inspection framework, secondarily to health inspectors within water protection area inner catchments (WA 2002: Articles 174 and 175).

Slovenia's legislation for inspection is complex and systemically regulated by the Inspection Act of 2002 (IA), containing procedural, organisational, and substantive provisions (Pirnat et al., 2004). Procedural issues are further regulated by sector-specific laws, laws regulating sector-specific inspection, and legislation regulating general administrative procedure (General Administrative Procedure Act, 1999 with amendments); there is a subsidiarity relationship amongst these laws (Kovač Kovač, 2016a, b). The IA applies to all administrative supervision areas, including water protection. In this context, the WA is *lex specialis* pursuant to its establishing special rules and powers for inspectors in the field of water protection.

Inspectors have special powers and responsibilities and may take special measures.[32] After procedure completion for establishing facts and taking evidence, an inspector may order preventive measures, regular remedial measures, special measures, and measures for the protection of the rights of other persons: the first set of measures includes verbally stating irregularities and consequences thereof; the second includes ordering measures by means of administrative decision; the third includes imposing temporary prohibition of activity performance; the fourth includes informing the public by publishing a decision in the media (Jerovšek & Kovač, 2008).

In addition to powers determined by the IA, inspectors also have the following special powers determined by the WA (2002, Article 175):

- to prohibit the use of water if water is used without water permit or concession, or not recorded wherever so prescribed;
- to prohibit and restrict the exercise of water rights if exercised contrary to a water permit or concession;

[31] In Slovenia, the legal status of inspectors is unique. Their powers are broader than regular civil servants, their independence particularly emphasised, as they, for example, may issue decisions in administrative proceedings directly on the basis of the law (Article 28 of the General Administrative Procedure Act, 1999 with amendments). Additionally, they may not, owing to work requirements, be transferred without their consent to another post that does not require the exercise of powers of inspection.

[32] E.g. unless otherwise provided by law, an inspector has the right to enter premises, buildings, and land and to access equipment and facilities without prior notification and without the permission of the person liable or the responsible person acting on behalf of the person liable and regardless of working hours. The person liable may deny an inspector entry to residential premises if the inspector does not present an appropriate decision issued by the competent court.

- to halt construction and other physical space intervention if executed without or contrary to water approval;
- to order temporary suspension of activities if provisions of the Act are violated;
- to prohibit the use of a facilities and installations if used without or contrary to water permit or concession;
- to prohibit and restrict substance discharge or drainage into waters if executed contrary to the provisions of the Act and to order restoration to a previous state;
- to prohibit the removal of alluvium from surface waters if taken contrary to the Act;
- to order damage remedy and the restoration to a previous state;
- to order other necessary measures in accordance with the Act or regulations issued on the basis thereof.

Under conditions specified in the WA, inspectors may issue oral decisions and immediately order implementation, such as when direct danger pollution of drinking water source exists (WA 2002: Article 176).

Appeals against inspectors' decisions are decided on by the ministry but do not stay decision execution (IA 2002: Article 30 in connection with the General Administrative Procedure Act 1999: 224; WA 2002: Article 175).

When inspectors detect offence, they may impose administrative fines and seize objects that have been used or intended for, or resulted from an offense (WA 2002: Article 181).

In practice, Slovenia faces capacity issues; Between 2005 and 2015, for example, the number of regulations increased sevenfold, whilst the number of inspectors remained unchanged (Ministry of Agriculture and the Environment, 2015), the Inspectorate for Environment and Spatial Planning's 2016 annual report stating, "[due] to insufficient staffing, inspectorate cannot provide systematic and planned inspection control in the field of water management and regulation" (Inspectorate for Environment and Spatial Planning, 2017).[33] The situation is even more worrying when we take into account issues related to water protection supervisors (*infra*, Chap. 4.3).

4.3 Water Protection Supervisors

The second type of institutionalised water supervision is *sui generis*, as defined by WA (2002: Articles 177 to 180).[34] This type of supervision is supplementary to inspection and carried out within the framework of services of general interest.

[33] Each territorial branch employs at least one inspector, specialised in water protection, but within whose competences also fall protection of environment and nature (Inspectorate for Environment and Spatial Planning, 2019).

[34] The same form of protection is principally provided by EPA. Environmental protection supervisors perform selected tasks, e. g. ascertaining facts and circumstances, and monitoring compliance with measures issued by inspectors, in the area of water pollution from small municipal wastewater treatment plants, and craft and craft-like activities. The difference is that they perform their tasks within inspection, not within providers of services of general interest (EPA 2004: Articles 158–160).

Namely, water protection supervision is carried out by river and marine supervisors (hereinafter: water protection supervisors), established by individual public service water regulation provider, who carry out work on the basis of employment and other appropriate contracts (*supra*, Fig. 9.3).

Water protection supervisors perform their competences on the basis of granted public authority (Constitution 1991: Article 121),[35] and are specially trained natural persons authorised by the environmental minister; they must also successfully qualify, by means of examination, to conduct minor offence procedures in accordance with the law regulating minor offences.[36]

The tasks of water protection supervisors comprise the following (WA 2002: Article 178–179):[37]

1. direct supervision of water condition and regime;
2. supervision of prohibition implementation;
3. determination of actual condition in the event of violation of prohibitions determined in the Act, in regulations issued on the basis thereof, and other environmental protection regulations applying to the burdening of waters; and
4. drawing the attention of various persons to prescribed action in relation to the burdening of waters.

When water protection supervisors establish that statutory provisions have been violated that exceed their competence, they must inform a competent inspector of said violation. If there are reasons for suspecting that a criminal offence has been committed for which prosecution is *ex officio*, water protection supervisors must secure the site where the alleged criminal offence occurred and evidence thereof, and immediately notify the police (WA 2002: Article 179).

If water inspection supervisors detect offences, they may impose administrative fines and seize objects that have been used or intended for, or resulted from an offense (WA 2002: Articles 179 and 181).

In practice, only approximately 20 water protection supervisors have already obtained the national vocational qualification, but since subsidiary legislation has

[35] Water protection supervisors, as with inspectors, are independent whilst carrying out water protection supervision, conducting procedure, and deciding on minor offences in minor offence procedures in accordance with the law regulating minor offences.

[36] The basis for obtaining the authority to carry out water protection supervision is the holding of a national vocational qualification certificate for "water protection supervisors" issued in accordance with regulations governing national vocational qualifications. Water protection supervisors carry an official badge and identity card. They identify themselves as officials by means of official identity card.

[37] Water protection supervisors have the right to undisturbed access to water or waterside land, and to enter water facilities and installations intended for the special use of water. They may furthermore require and receive from owners, other users, or managers of water facilities and installations access to the identification documents of persons and documents on acquired water rights and recorded special water uses. They may also document any violations of the provisions of the WA and of regulations issued on the basis thereof.

yet to be adopted,[38] they do not have full power to act (Inspectorate for Environment and Spatial Planning, 2017, 2018, 2019). Consequently, if they detect irregularities, they notice the inspectorate, which is, as aforementioned, understaffed. Additionally, a mechanism for coordinating the work of water protection supervisors and inspectors competent for agriculture will have to be established as this is not systematically regulated in the WA (Ministry of Agriculture and the Environment, 2015).

4.4 Voluntary Water Protection Supervisors

WA's amendment in 2008 provided for a new form of water protection: voluntary water protection supervisors;[39] voluntary water protection supervisor may be any (1) natural person of full age who (2) is a national of the Republic of Slovenia (3) meeting prescribed conditions for water protection supervisors, (4) who has obtained written authorization from a minister, (5) who register themselves with the public service provider for the area they wish to supervise (WA 2002, Article 180.a).

Voluntary water protection supervisors are only authorised to directly monitor water condition and regime, and to warn people of prescribed behaviour in relation to water pollution. Their competence is thus limited when compared to water protection supervisors. They must immediately notify water protection supervisors or police if they suspect that minor or criminal offence in relation to waters has been committed; in doing so, they must evidence the facts and hand over any evidence obtained in relation to offence and secure the crime scene (WA 2002: Articles 179, 180 and 180.a).

5 Conclusion

Slovenia is one of the few countries in the world which regulates the right to drinking water at a constitutional level, definitely evidencing the high level of importance this right has for Slovene society; the constitution imposes obligations on the state to provide conditions to exercise this right. However, as argued above, despite the right to water being defined as a human right, this provision does not oblige the state to provide water supply in areas where only self-sufficiency is rational or if the real estate cannot be connected to water infrastructure. Moreover, 2016's amendment to

[38] Regulation, specifying the detailed tasks of water protection supervisors and setting the methods and conditions for the organisational, substantive and territorially harmonised implementation of water protection supervision is crucial (WA 2002: Article 178, Paragraph 2).

[39] The same form of protection is envisaged in the Nature Conservation Act of 1999 (Article 159).

the constitution affirms that water resources serve primarily and sustainably for the drinking water supply of population and households and are, in this respect, not classified as tradable goods. Consequently, water resources are a public good managed, not owned, by the state, and this means all natural and artificial water resources. Additionally, the state is responsible for protecting water resources, such as monitoring water status, preventing and reducing pollution, encouraging sustainable water use, and protecting water environment. In accordance with the constitutional principle of free economic initiative, amendment 2016 still enables the exploitation of water resources for special use, such as the production of beverages, the state conferring water permits and concessions for such use. However, drinking water supply takes precedence, so such exploitation shall not endanger water resources. Moreover, according to 2016's amendment, drinking water supply is no longer an obligatory local public service; on the contrary, the state is deemed responsible for the water resource management, supplying drinking, direct household and non-profit user supply through local self-government communities. However, this new separation of competence between state and municipalities lacks clarity and requires further interpretation by means of new sector specific legislation, still to be adopted despite its May 2018 deadline. Such delay in harmonising sector specific legislation with 2016's constitutional amendment is a breach of Article 153 of the Constitution, which requires legal act conformity. Finally, as this right is rather new, the Constitutional Court has yet to establish practice on it.

The institutional setting of water governance is complex, including private and public law bodies, central, regional and local levels of governance, and institutionalised and non-institutionalised organisations, making coordination difficult, especially taking into account the absence of (strong) centre of government (Rakar & Virant, 2019). Water regulation legislative techniques are similarly complex, encompassing laws and contracts, and public law and private law forms of (legal) acts. However, legislation implementation, especially supervision, faces challenges in relation to human resources management and regulatory activities. Namely, despite evident irregularity and the existence of vocational qualifications, the government has not adopted secondary legislation, leaving water protection supervisors much less effective. Similarly, the number of employed inspectors has not sufficiently grown in accord with growth in environmental legislation, making supervision prioritisation inevitable and increasing the potential for unsanctioned behaviour.

References

Akhmouch, A., & Correia, F. N. (2016). The 12 OECD principles on water governance—When science meets policy. *Utilities Policy, 43*(A), 14–20. https://doi.org/10.1016/j.jup.2016.06.004

Avbelj, M. (2019). *Komentar Ustave Republike Slovenije*. Nova Gorica: Evropska pravna fakulteta.

Berginc, M., Kremesec-Jevšenak, J., & Vidic, J. (2007). *Sistem varstva narave v Sloveniji*. Ljubljana: Ministrstvo za okolje in prostor.

Biermann, B. (2007). Umwelt. In A. Benz, S. Lütz, U. Schimank, & G. Simonis (Eds.), *Handbuch governance* (pp. 424–436). Wiesbaden: VS Verlag für Sozialwissenschaften.

Brilly, M. (2000). Koncept in izhodišča Zakona o vodah ter načrtovanje v vodnem gospodarstvu. *Vodno-gospodarski biro Maribor*. Retrieved 6 September, 2019, from http://mvd20.com/LETO2000/R13.pdf

Briscoe, J. (2011). Making reform happen in water policy: Reflections from a practitioner. *Background Paper, OECD Global Forum on Environment: Making Water Reform Happen, 25–26 October*. Paris: Organization for Economic Cooperation and Development.

Civil Servants Act (CSA). (2002). *Official Gazette of the Republic of Slovenia* (56/2002 and amendments).

Council of Europe. (2001). Recommendation Rec (2001)14 of the Committee of Ministers to member states on the European Charter on Water Resources. Retrieved 19 May, 2020, from https://rm.coe.int/1680504d85

Council of Europe. (2008). *The Strategy for Innovation and Good Governance at Local Level.* Retrieved 13 March, 2018, from https://rm.coe.int/1680701699

Constitution of the Republic of Slovenia. (1991). *Official Gazette of the Republic of Slovenia* (33/91-I and amendments).

Dupuy, P.-M., & Viñuales, J. E. (2016). *International environmental law*. Cambridge: Cambridge University Press.

Environmental Protection Act (EPA). (2004). *Official Gazette of the Republic of Slovenia* (41/04 and amendments).

European Convention on Human Rights (ECHR). (1950). Retrieved 19 May, 2020, from https://www.echr.coe.int/Pages/home.aspx?p=basictexts

Gantar, P. (2016). *Ustava in voda: za ponovni premislek.* Dnevnik. https://www.dnevnik.si/1042754757

General Comment No. 15 to the ICESR. (2002). General Comment No. 15: The Right to Water (Arts. 11 and 12 of the Covenant) Adopted at the Twenty-ninth Session of the Committee on Economic, Social and Cultural Rights, on 20 January, 2003 (Contained in Document E/C.12/2002/11). Retrieved 12 March, 2018, from http://www.refworld.org/pdfid/4538838d11.pdf

General Administrative Procedure Act (GAPA). (1999). *Official Gazette of the Republic of Slovenia* (80/99 and amendments).

Godden, L., & Ison, R. (2019). Community participation: Exploring legitimacy in socio-ecological systems for environmental water governance. *Australasian Journal of Water Resources, 23*(1), 45–57. https://doi.org/10.1080/13241583.2019.1608688

Government of the Republic of Slovenia. (2012). *Predlog zakona o spremembah in dopolnitvah Zakona o vodah* (27. junij 2012, št. 2012-2330-0179). Retrieved 13 March, 2018, from http://imss.dz-rs.si/imis/ca119e8e94a78169d929.pdf

Inspectorate for Environment and Spatial Planning. (2017). *Poročilo o delu za leto 2016.* Retrieved 16 March, 2018, from http://www.iop.gov.si/fileadmin/iop.gov.si/pageuploads/5_O_INSPEKTORATU/Porocila_in_nacrti_dela/Porocilo_IRSOP_2016.pdf

Inspectorate for Environment and Spatial Planning. (2018). *Poročilo o delu za leto 2017.* Retrieved 8 November, 2019, from http://iop.arhiv-spletisc.gov.si/fileadmin/iop.gov.si/pageuploads/5_O_INSPEKTORATU/Porocila_in_nacrti_dela/Porocilo_IRSOP_2017.pdf

Inspectorate for Environment and Spatial Planning. (2019). *Poročilo o delu za leto 2018.* Retrieved 8 November, 2019, from https://www.gov.si/assets/organi-v-sestavi/IRSOP/O-IRSOP-u/porocila-o-delu/POROCILO-O-DELU-IRSOP-2018.pdf

Inspection Act (IA). (2002). *Official Gazette of the Republic of Slovenia* (56/2002 and amendments).

International Covenant on Economic, Social and Cultural Rights (ICESR). (1966). Retrieved 12 March, 2018, from http://www.ohchr.org/EN/ProfessionalInterest/Pages/CESCR.aspx

Jerovšek, T., & Kovač, P. (2008). *Posebni upravni postopki.* Ljubljana: Fakulteta za upravo.

Knez, R. (2015a). *Is it necessary to regulate a right to drinking water in the constitution*? University of Maribor, Jean Monnet Center of Excellence. Retrieved 12 March, 2018, from http://www.jm-excellence.si/is-it-necessary-to-regulate-a-right-to-drinking-water-in-the-constitution/

Knez, R. (2015b). *Export of drinking water by public companies*. University of Maribor, Jean Monnet Center of Excellence. Retrieved 12 March, 2018, from http://www.jm-excellence.si/export-of-drinking-water-by-public-companies/

Köck, W. (2012). Pläne. In W. Hoffmann-Riem, E. Eberhard Schmidt-Aßmann, & A. Voßkuhle (Eds.), *Grundlagen des Verwaltungsrechts, Band II (Informationsordnung, Verwaltungsverfahren, Handlungsformen)* (pp. 1389–1456). München: C. H. Beck.

Kovač, P. (2016a). Inspection regulation between general procedural codification and field specifics—A case study of Slovenia. *Danube: Law and Economics Review, 7*(1), 1–17.

Kovač, P. (Ed.). (2016b). *Inšpekcijski nadzor*. Ljubljana: Uradni list Republike Slovenije.

Kovač, P. (2017). Between theoretical principles and practice in Slovene regulatory impact assessment procedures. *Review of Central and East European Law, 42*(2–3), 215–250.

Kralj, M. (2017). *Voda bo morda kriva za spremembo več kot ducat zakonov*. Retrieved 12 March, 2018, from https://www.dnevnik.si/1042787174/slovenija/voda-bo-morda-kriva-za-spremembo-vec-kot-ducat-zakonov

Lex Localis. (2018). *Katalog pristojnosti občin*. Retrieved 14 March, 2018, from http://www.lex-localis.info/TabView/VsebinaTabs.aspx?SectionID=c6c26383-9df0-4f89-8f05-67d774f72bab

Mechlem, K. (2016). Groundwater governance: The role of legal frameworks at the local and national level—Established practice and emerging trends. *Water, 8*(8), 347–362. https://doi.org/10.3390/w8080347

Ministry of Agriculture and the Environment. (2015). *Poročilo o izvajanju programa ukrepov upravljanja voda za obdobje 2011–2015*. Ljubljana: Ministry of Agriculture and the Environment.

Ministry of Environment and Spatial Planning. (2018). *Water*. Retrieved 8 November, 2019, from http://www.mop.gov.si/si/medijsko_sredisce/novica/8394/

Nanni, M. (2007). *Principles of water law and administration*. London: CRC Press.

Nature Conservation Act (NCA). (1999). *Official Gazette of the Republic of Slovenia* (56/99 and amendments).

OECD. (2015a). *OECD Principles on Water Governance*. Retrieved 13 March, 2018, from http://www.oecd.org/governance/oecd-principles-on-water-governance.htm

OECD. (2015b). *OECD Regulatory Policy Outlook 2015*. Country profile Slovenia. Retrieved 13 March, 2018, from http://www.oecd.org/gov/regulatory-policy/Slovenia-web.pdf

OECD. (2015c). *Water resources allocation: Sharing risks and opportunities*. Paris: OECD Publishing. https://doi.org/10.1787/9789264229631-en

OECD. (2018). *Regulatory policy in Slovenia: Oversight matters*. Paris: OECD Publishing. https://doi.org/10.1787/9789264291690-en

Ozerol, G., Vinke-de-Kruijf, J., Brisbois, M., Casiano Flores, C., Deekshit, P., Girard, C., et al. (2018). Comparative studies on water governance: A systematic review. *Ecology and Society, 23*(4), 1–28.

Parliamentary Assembly of the Council of Europe. (2009). Resolution 1693. Retrieved 19 May, 2020, from http://assembly.coe.int/nw/xml/XRef/Xref-XML2HTML-EN.asp?fileid=17786&lang=en

Pavčnik, M. (2015). *Teorija prava* (5th ed.). Ljubljana: GV založba.

Pečarič, M. (2016). *Regulativne tehnike*. Ljubljana: Uradni list Republike Slovenije.

Pekolj, N. (2014). Ko kozarec pitne vode določi meje notranjega trga EU. *Pravna praksa, 33*(26), 11–13.

Pevcin, P., & Rakar, I. (2018). Evaluation of delivery mechanisms in the water supply industry–Evidence from Slovenia. In I. Koprić, H. Wollmann, & G. Marcou (Eds.), *Evaluating reforms of local public and social services in Europe, more evidence for better results* (pp. 207–223). Cham: Palgrave Macmillan.

Pirnat, R., Bugarič, B., Jerovšek, T.,Herševan, E., Pilčanič, S., Purg, K. Š., et al. (2004). *Komentar zakonov s področja uprave*. Ljubljana: Inštitut za javno upravo pri Pravni fakulteti.

Pličanič, S. (2017). Pravni vidiki trajnostnega upravljanja voda (aktualna vprašanja). *Vodni dnevi 2017* (pp. 37–45), Portorož 5.

Rakar, I. (2011). *Vključevanje javnosti v delovanje izvršilne veje oblasti pri sprejemanju splošnih pravnih aktov v Republiki Sloveniji: doktorska disertacija*. Ljubljana: Pravna fakulteta.

Rakar, I., & Kovač, P. (2019). Boljši predpisi kot horizontalna politika razvoja slovenske javne uprave: med cilji in realnostjo. In M. Pečarič (Ed.), *Od boljših predpisov k njihovemu učinkovitejšemu izvrševanju* (pp. 45–64). Ljubljana: Fakulteta za upravo.

Rakar, I., & Tičar, B. (2017). *Pravo javne uprave*. Ljubljana: Fakulteta za upravo.

Rakar, I., Tičar, B., & Sever, T. (2017). Public participation in rulemaking—A form of legal protection? In A. M. Kruglashov (Ed.), *State and society: tools, mechanisms and principles of interaction [Держава і суспільство: інструменти, механізми та принципи взаємодії]* (pp. 14–35). Chernivtsi: Bukrek.

Rakar, I., & Virant, G. (2019). Public administration characteristics in Slovenia. In *Unpublished report of the EUPACK2019 project*. Ljubljana: Faculty of Public Administration.

Report on the Preparation of Constitutional Act on Supplementing Chapter III of the Slovene Constitution. (2016). Retrieved 12 March, 2018, from http://imss.dz-rs.si/imis/295c7356c294eebf4b74.pdf

Rules on determining and classification for water bodies on surface water. (2005). *Official Gazette of the Republic of Slovenia* (63/2005 and amendments).

Sancin, V., Pucelj Vidovič, T., Kovič Dine, M., & Vrbica, S. (Eds.). (2015). *Odgovornost države zagotavljati pravico do čiste pitne vode*. Pravna klinika za varstvo okolja 2014/2015. Ljubljana: Pravna fakulteta, Univerza v Ljubljani.

Schophaus, M. (2019). Umweltgovernance und Governance für eine nachhaltige Gesellschaftstransformation. In K. Möltgen-Sicking & T. Winter (Eds.), *Governance: Eine Einführung in Grundlagen und Politikfelder* (pp. 129–148). Wiesbaden: Springer.

Sever, T., Rakar, I., & Kovač, P. (2014). Protecting human rights through fundamental principles of administrative procedures in Eastern Europe. *Danube: Law and Economics Review, 5*(4), 249–275. https://doi.org/10.2478/DANB-2014-0014

Šturm, L. (Ed.). (2011). *Komentar Ustave Republike Slovenije*. Brdo: Fakulteta za državne in evropske študije.

Thielbörger, P. (2014). *The right(s) to water, the multi-level governance of a unique human right*. Berlin Heidelberg: Springer.

Tičar, B., & Rakar, I. (2011). *Pravo javnega sektorja*. Maribor: Inštitut za lokalno samoupravo in javna naročila.

UN General Assembly. (1979). Convention on the elimination of all forms of discrimination against women. *United Nations*. Retrieved 12 March, 2018, from http://www.ohchr.org/Documents/ProfessionalInterest/cedaw.pdf

UN General Assembly Resolution /64/292. (2010). *The human right to water and sanitation*. Retrieved 12 March, 2018, from http://www.un.org/es/comun/docs/?symbol=A/RES/64/292&lang=E

UNESCO. (2012). *Vidik človekovih pravic glede naravne in kulturne dediščine: Vloga UNESCO*. Retrieved 12 March, 2018, from http://imss.dz-rs.si/imis/295c7356c294eebf4b74.pdf http://www.skupaj-mocnejsi.si/wp-content/uploads/2013/06/Vidik-Clovekovih-Pravic-UNESCO.pdf

UNESCO. (2015). *Water for a sustainable world*. Paris: UNESCO.

United Nations Water Conference. (1977). Retrieved 19 May, 2020, from https://www.ircwash.org/sites/default/files/71UN77-161.6.pdf

van Kralingen, R. W. (1997). A Conceptual Frame-based Ontology for the Law. In Proceedings of the First International Workshop on Legal Ontologies (LEGONT '97) (pp. 15–22). Melbourne (Australia): University of Melbourne.

Virant, G., & Rakar, I. (2017). *Report on public administration reform trends and reform dynamics in Slovenia*. Brussels: European Commission.
Virant, G., & Rakar, I. (2018). *Public administration characteristics and performance in EU28: Slovenia*. Brussels: European Commission. Retrieved 8 November, 2019, from https://ec.europa.eu/social/main.jsp?catId=738&langId=en&pubId=8123&furtherPubs=yes
Waters Act (WA). (2002). *Official Gazette of the Republic of Slovenia* (67/02 and amendments).
Woodhouse, P., & Muller, M. (2017). Water governance—An historical perspective on current debates. *World Development, 92*, 225–241. https://doi.org/10.1016/j.worlddev.2016.11.014

Chapter 10
Water Crime in the Republic of Slovenia

Saša Kuhar and Gorazd Meško

1 The Republic of Slovenia

The Republic of Slovenia (hereinafter Slovenia), a Central European country, is the former Republic of Yugoslavia. On 25 June 1991, Slovenia officially declared its independence and became a Democratic parliamentary republic. Slovenia became a member of the European Union on May 1, 2004. Slovenia has a population of 2.1 million, of which 83.1% (2002 census) are Slovenes. Population density is 101.2 inhabitants per km^2 (Government of the Republic of Slovenia, n.d.) which is lower than in the majority of other European states. In Slovenia, there are also two national minority communities of Italians and Hungarians. They are considered indigenous minorities, and their rights are protected under the Constitution. Similarly to other modern societies, the country has been facing demographic issues, such as ageing, and low birth rate. According to the Government of the Republic of Slovenia (n.d.) Slovenia is approximately 50% urban and 50% rural.

Slovenia has a quite stable political system characterised by smooth transfers of power and regular elections. The state's authority is based on the principle of the separation of legislative, executive and judicial powers, with a parliamentary system of government. The highest legislative authority is the National Assembly, which has the right to enact laws. The Government of the Republic of Slovenia exercises executive power in Slovenia according to the Constitution and the laws of Slovenia. It is also the highest administrative authority in Slovenia. The government carries out the country's domestic and foreign policy, shaped by the National Assembly. It directs and coordinates the work of government institutions and bears full

S. Kuhar (✉)
Slovenian Nuclear Safety Administration, Ljubljana, Slovenia
e-mail: sasa.kuhar@gov.si

G. Meško
Faculty of Criminal Justice and Security, University of Maribor, Ljubljana, Slovenia
e-mail: gorazd.mesko@fvv.uni-mb.si

K. Eman et al. (eds.), *Water, Governance, and Crime Issues*,
https://doi.org/10.1007/978-3-030-44798-4_10

responsibility for everything occurring within the authority of executive power. The government, headed by the Prime Minister, thus represents the political leadership of the country and makes decisions in the name of the executive power. Political stability index from the World Bank was 0.99[1] in the year 2016 (The Global Economy, n.d.). The Best Country Risk Report (2017) reported that the GDP per capita in Slovenia was 21,320 USD in the year 2017.

OECD (2017) states that 87% of adults aged 25–64 have completed upper secondary education in Slovenia. The education level is higher than the OECD average, which is 74%. Concerning the quality of the educational system, the average student scored 509 in mathematics, reading literacy and science in the OECD's Program for International Student Assessment (PISA), slightly more than the OECD average of 486 points. On average, girls outperformed boys' by15 points, higher than the average OECD gap of 2 points. According to the UN Human Development Report (UNDP, 2015), Slovenia remains in the group of countries with very high human development. The human development index (HDI 0.880) ranks Slovenia 25th out of 188 countries, which is the same place as a year earlier (out of 187 countries). The average household net-adjusted disposable income per capita is USD 20,505 a year, which is lower than the OECD average of USD 30,563. Household financial wealth is the total value of a household's financial worth, such as money or shares held in bank accounts. In Slovenia, the average household net financial wealth per capita is estimated at USD 20,048, much lower than the OECD average of USD 90,570. These data do not include non-financial assets (e.g., land and dwellings). There is a considerable gap between the richest and most deprived in the country. The top 20% of the population earn close to four times as much as the bottom 20%. Slovenians earn USD 34,965 per year on average. That is less than the OECD average (USD 44,290) (OECD, 2017). According to the Eurostat (2016), women in Slovenia earned on average 7.8% less than men in 2016.

1.1 Geographic Background

The Republic of Slovenia lies at the heart of Europe, in Central and Southeastern Europe. Four major European geographic regions meet in Slovenia: The Alps, the Dinarides (also known as Dinaric Alps), the Pannonian Plain, and the Mediterranean. Although on the shore of the Adriatic Sea near the Mediterranean Sea, most of Slovenia is in the Black Sea drainage basin. The Alps (the Julian Alps, the Karavanke chain, the Kamnik-Savinja Alps and the Pohorje massif) dominate Northern Slovenia along its long border with Austria. Slovenia's Adriatic coastline stretches approximately 47 km from Italy to Croatia. The Karst Plateau is located in the southwest. A limestone region of underground rivers, gorges, and caves lies between Ljubljana and the Mediterranean. On the Pannonian plain to the East and Northeast,

[1] Political stability index is measured on a scale from 2.5 meaning weak to 2.5 meaning strong.

toward the Croatian and Hungarian borders, the landscape is mostly flat (Slovenia.si, n.d.). Protected areas of Slovenia include national, regional and nature parks, the largest of which is Triglav National Park. There are 286 Natura 2000 designated protected areas, which comprise 36% of the country's land area, the most significant percentage among European Union states.

According to the Ministry of the Environment and Spatial Planning (1998), there are 59 major rivers in Slovenia, altogether measuring about 2,500 km^2 in length. The total length of rivers in Slovenia is 26,989 km. The territory of Slovenia in 81% (16,423 km^2) belongs to the Black Sea basin and a smaller part 19% (16,423 km^2) to the Adriatic Sea basin. These two parts are divided into smaller units concerning their central rivers, the Mura River basin, the Drava River basin, the Sava River basin with Kolpa River basin, and the basin of the Adriatic rivers.

According to the Environmental Agency of the Republic of Slovenia (ARSO, n.d.) Slovenia has plentiful water resources, although they are not spatially uniformly distributed. Water covers about 272 km^2 of Slovenia's territory, while geologic units, which may store and transmit groundwater, contain about 50 m^3/s of the dynamic supply.

1.2 Environmental Context

Slovenia has a wealth of natural resources. The use of natural resources was increasing in the last years. The exploitation of natural resources generated or extracted in Slovenia has declined since 2007. According to the European Environment Agency (EEA, 2015), 22 million tons of natural resources were exploited in 2012. The vast majority were mineral resources used in the construction industry (lignite, lead, zinc, building stones, etc.). Imported materials (non-ferrous metals, petroleum, iron, steel, etc.) and goods also declined after 2008. Domestic material consumption per capita was 12.4 tons in 2012. The trend has changed in the year 2015; material productivity has been rising since 2007. This rise is one of the reasons that we increased recycling of the waste in recent years, and specific natural resources are returned to the production process. According to the Slovenian Statistical Office (2018), Slovenia's trade in goods continued to grow in 2017. Exports in 2017 were 13.1% higher than exports in 2016 (EUR 28.2 billion), while imports were 14.2% higher (EUR 27.5 billion).

Within the environmental and spatial policy, sustainable management of natural resources is formulated. The measures that it contains are still focusing mainly on waste management, wastewater, drinking water and water management, and construction of related infrastructure. The intention is that in the future, measures which focus on resources efficiency would be introduced in the context of this policy as well as they will be integrated into other sectoral policies.

Slovenia is rich in water, but water resources are vulnerable and distributed unevenly. Slovenia uses only a small portion of available water run-off for human use (about 3%). According to the EEA (2015), approximately 15 billion m^3 of water flows out of Slovenia through rivers every year, but long-term trends show that this volume is falling. Because of the country's headwater position and the torrential character of the majority of watercourses, the flow rate varies significantly during the year, with a lack of water in the vegetative period and large quantities of water (or even floods) during the other periods of the year. The annual groundwater recharge also varies considerably. Groundwater is a crucial resource in Slovenia. More than 97% of all drinking water is abstracted from groundwater bodies. Groundwater in Slovenia has all the crucial roles listed in the Water Framework Directive. Most groundwater bodies had a good chemical status.

Slovenia has an enormous advantage concerning the water reserves. According to the Environmental Agency of the Republic of Slovenia (Agencija Republike Slovenije za okolje [ARSO], 2015) in 2013 Slovenia had 2500 m^3 of groundwater per capita available. Thus, 3.6% of those volumes were in the same year pumped from aquifers mainly for drinking water supply, manufacture of beverages and SPA resorts. According to estimates of the ARSO the quantitative status of Slovenian groundwater is generally favourable, except in Murska basin, the groundwater level is gradually reduced (Eman, 2016). These data are confirmed by ARSO (2015), which is particularly emphasised in the Report on the Quantity of Groundwater in Slovenia.

Slovenia is one of the smallest countries in Europe, but concerning water resources, it is one of the richest according to the data (ARSO, n.d.) that the average annual precipitation that supplies surface waters and groundwater amount approximately 1500 mm. The mountainous western region receives the highest annual amount of precipitation, and the smallest amount occurs in the eastern part of Slovenia. ARSO (n.d.) stressed that there is a dense river network in Slovenia, but rivers are mostly short, with only 22% of rivers being longer than 25 km. Water quality in Slovenia is among the highest in Europe. One of the reasons is that most of the rivers rise on the territory of Slovenia. It does not mean that Slovenia has no problems with surface water and groundwater quality.

Some river sections are still loaded with excessive amounts of industrial and municipal wastewaters and are therefore substantially polluted or even overpolluted. Problems are also present in groundwater, which represents the primary source of drinking water in Slovenia. Groundwater is polluted with nitrates and pesticides, and, on a local level, additionally with chlorinated organic solvents. The highest level of pollution is registered in the north-eastern part of Slovenia (Mura and Drava basins) and near Celje (Nacionalni laboratorij za okolje in hrano, 2016).

Water Framework Directive (Direktiva Evropskega parlamenta in Sveta, 2000), which represents a framework for sustainable water management, set the goal that, by the year 2015, functional status of all waters would be achieved (WISE, 2008). United Nations (2017) set up 17 goals to transform our world. The sixth Sustainable Development Goal is to ensure access to water and sanitation for all. The main reason for this goal was that by the year 2050 "at least one in four people is likely to live in a country affected by chronic or recurring shortages of fresh water" (United Nations, 2017). Their goal is to access water for all people in the world. They believe that there is enough fresh water on the planet to achieve this goal. However, due to poor infrastructure and bad economics every year, millions of people, most of them children, die from diseases associated with inadequate water supply, sanitation, and hygiene. Brumen (2017) stressed that the Republic of Slovenia had to prepare two water governance plans thus far for the river basin districts of the Danube and the Adriatic Sea because it did not achieve its goals in the first period (by the year 2015). The first national water management plan was related to the period from 2009 to 2015, while the second water management plan is related to the period from 2015 to 2021.

1.3 The Legal Framework on Water and Water Management

The management of water resources in Slovenia is regulated by the Water Framework Directive (Direktiva Evropskega parlamenta in Sveta, 2000). The priority is to eliminate adverse effects on water, to ensure adequate water quality for humans and natural ecosystems and to preserve biodiversity. The first classification of surface water intended for the abstraction of drinking water in Slovenia was prepared in 2002 and published in the Official Gazette of the Republic of Slovenia. Furthermore, in 2002, the Water Act (Zakon o vodah [ZV], 2002) defined water as a natural public asset. Since the water is a public good managed by the state (ARSO, 2008).

The general legislation and regulatory framework for water management in Slovenia is the Water Act (hereinafter ZV). Additionally, some other regulations are essential for the protection of water, the quality of surface and groundwater in Slovenia: (1) Decree on groundwater status (Uredba o stanju podzemnih voda, 2009);[2] (2) Rules on drinking water (Pravilnik o pitni vodi, 2004);[3] and (3) Decree on surface water status (Uredba o stanju površinskih voda, 2009).[4] The Water Act (2002) established protected areas where drinking water is significant. Furthermore, Slovenian legislation has two legal acts related to the water supply. Water Framework Directive (Direktiva Evropskega parlamenta in Sveta, 2000) aims to ensure the excellent chemical status of both surface water and groundwater bodies. It is implemented in the Water Act (ZV, 2002). Environment Protection Act (Zakon o varstvu okolja [ZVO-1], 2006) including all belonging acts. The Water Act (2002) deals with the ownership, control, and use of water as a resource. Water Act also governs public assets and public services in the area of waters, water facilities and installations, and other water-related issues.

Decree on drinking water supply (Uredba o oskrbi s pitno vodo, 2012) establishes the tasks that are carried out within the framework of mandatory municipal

[2] Before the year 2009, it was Decree on the quality of underground water (Uredba o kakovosti podzemne vode 2002).

[3] Before the year 2004, it was Rule on drinking water quality (Pravilnik o zdravstveni ustreznosti pitne vode 1997).

[4] Before the year 2009, it was Decree on the chemical status of surface waters (Uredba o kemijskem stanju površinskih voda 2002).

utility supply of drinking water (hereinafter referred to as public service), and specific conditions for the supply of drinking water, which is implemented as a public service, and for their own care drinking water. This regulation, as mentioned above, lays down standards of public utilities, which must be met for public service and lays down measures for the public service realisation. Thus, the drinking water supply operators are obliged to ensure the quality of drinking water, but they are not responsible for the state of installations at the end-user location.

Rules on drinking water supply (Pravilnik o oskrbi s pitno vodo, 2006) determine the duties that have to be carried out to run the public drinking water service. It has to be organised by the local community in its area with exception when the elevation is above 1500 m or in areas where less than 50 residents live. For the use of the surface water or groundwater for public drinking supply, the water permission has to be received. For this, the municipality has to obey the rules and standards, in which public service is determined (Eman, Kuhar, & Meško, 2016).

The Environmental Protection Act (Zakon o varstvu okolja [ZVO-1], 2006) governs the protection of the environment against burdens. Protection of the environment is a prerequisite for sustainable development, and in this context defines the basic principles of environmental protection, environmental protection measures, environmental monitoring and environmental information, economic and financial instruments for environmental protection, public services for environmental protection and other related issues (Banjanac, 2013).

The protection of drinking water resources is a subject of the regulation by municipalities and the Ministry of the Environment and Spatial Planning [Ministrstvo za okolje in prostor]. Considering the way of the decision-making, according to Eman et al. (2016), all drinking water resources of Slovenia will be protected by governmental decrees in the next few decades. In the process of adopting the decrees, the drinking water supply operators are not a party to the procedure, nor are they invited to offer their opinion refer the planned activities in the drinking water protection zones. The legal regulation of water supply is recommended because such measures can reduce illegal exploitation of water resources.

The Slovenian Water Agency [Direkcija za vode] started operating on January 1, 2016, as a new body affiliated to the Ministry of the Environment and Spatial Planning. Establishment of this new body has led to an elaborate reorganisation in the field of implementation of the water management process. The reason was to follow the objective of the provision of efficient and professional planning and management of waters. The objective of the establishment of the Agency is "to establish a system of integral water management: to utilize the water potential of Slovenia as a development opportunity, take into account the water capacity in spatial planning and reduce risks to life, health and property of residents" (Direkcija Republike Slovenije za vode, 2016).

The activities of the Environmental Agency of the Republic of Slovenia in the field of water are:

- preparing programs for monitoring the quality of waters (rivers, lakes, groundwater, and sea),

- determining water pollution by physiochemical and biological analyses,
- monitoring and measuring individual elements of the water cycle at hydrologic monitoring stations for surface water (watercourses, lakes, sea) and groundwater and springs,
- measuring levels and temperature of groundwater, springs, rivers, lakes and sea, and determines changes,
- determining and monitoring river discharge regimes, and notes changes based on water level and hydrometric measurements,
- issuing warnings of an increased risk of flooding from rivers and sea, and of declining water levels of surface and groundwater,
- managing waters in a manner to protect water in all its forms regulates waters and decides on the use of water by the following principles:
  - the integrity of river basin, considering natural processes, water dynamics, interaction and interdependence of water and terrestrial ecosystems,
  - ensuring protection against adverse effects of waters, which arise from the security needs of the population and its property,
  - preserving and regulating the water quantity,
  - achieving good water status in terms of quality,
  - long-term protection of available water resources,
  - encouraging the sustainable use of water,
  - an economic evaluation of water (including the costs of water burdening, protection and regulation in the price of water),
  - public participation in adopting water management plans,
  - considering the best available techniques and scientific findings relating to the principles of nature—all these are realistic goals of water management;
- conducting administrative procedures about water quality, and issuing administrative decisions based on regulations dealing with emissions into water and soil,
- conducting administrative procedures and issuing administrative decisions on the assessment of environmental taxes and exemption from the payment of environmental taxes with regard to emissions into water (this procedure also involves monitoring payments and administrative execution),
- conducting procedures, keeping the account of water fees and monitoring payments,
- drawing up operational programs for reducing emissions into water and soil,
- directing and monitoring the implementation of rehabilitation plans,
- issuing authorisations for monitoring water and soil,
- keeping databases, recording, and registering in the area of water quality and quantity,
- monitoring and processing reports on wastewater monitoring,
- providing the expert basis for conducting administrative procedures and preparing regulations in the field of emissions into water and soil,
- preparing implementation programs in the area of transposition of the EU acquis,
- performing professional tasks to implement international obligations (ARSO, n.d.).

1.4 The Water Governance in Slovenia

Water as a source is a natural public good and a public property in the Republic of Slovenia. The Republic of Slovenia was the first country that adopted a human right to water[5] into a constitution in the year 2016. The reason why it is essential to put the right to safe drinking water into the constitution is that water is a vital part of the environment, an essential commodity and by its nature a given resource, which is of strategic importance to human life, development, prosperity and the security of the country. This constitutional change requires that the government shall take care of safe drinking water in the Republic of Slovenia. Drinking water resources and supplying the population with the water will have a priority over economic use, such as bottled water and selling beverages. Water is becoming a critical raw material of the twenty-first century. The water consumption in the world will increase as a result of the rise in population and rising standards in developing countries. The water supply will become more unpredictable, also due to the impact of climate change. An excellent example of water scarcity is the Cape Town, where at the beginning of the year 2018, the city remained entirely without the water due to the worst drought in the history of the country (Welch, 2018). From this point of view, the adoption of a human right to drinking water into a constitution appears to be a smart and necessary step. It is, therefore, of the crucial importance of how natural resources are managed and by whom.

The water governance in Slovenia is the responsibility of the state. After the Republic of Slovenia ratified the Aarhus Convention (Zakon o ratifikaciji Konvencije o dostopu do informacij, udeležbi javnosti pri določanju in dostopu do pravnega varstva v okoljskih zadevah, 2004) a request was triggered by the political system regarding a change in the way the political institution's function. According to Brumen (2017), the inclusion of the civil society into the processes of decision making and implementation in the field of the environment protection has thus become a legal norm. Public participation is also a part of several other Slovenian normative acts, which manage environmental protection in the Republic of Slovenia. The principle of public participation in the Water Act (ZV, 2002) is counted among six fundamental principles of water governance, defined in Article three of the Water Act (ZV, 2002) along with Article 58, which defines public participation in the adoption of water governance plans in more detail. The Public also had an essential role in the process of entering the right to drinking water into the constitution of the Republic of Slovenia.

[5] Article 70a (Right to Drinking Water): "Everyone has the right to drinking water. Water resources shall be a public good managed by the state. As a priority and in a sustainable manner, water resources shall be used to supply the population with drinking water and water for household use and in this respect shall not be a market commodity. The supply of the population with drinking water and water for household use shall be ensured by the state directly through self-governing local communities and on a not-for-profit basis" (Constitution of the Republic of Slovenia 1991: Amendment 2016).

Drinking water supply in Slovenia is the responsibility of individual municipalities, which need to provide the necessary amounts of qualitative drinking water to their citizens. According to the Statistical Office of the Republic of Slovenia (Statistični urad Republike Slovenije, 2015), 77% of the drinking water supply of the population is organised through public networks, 14% from private wells, 5% from rainwater reservoirs and 4% from other sources (Fig. 10.1).

Training programs for local-level, monitoring and institutional organisation to prepare, manage and supervise the implementation of water-supply projects, wastewater collection, and treatment within the river basins are provided by National Environment Protection Action Program (Nacionalni program varstva okolja, 1999). In the sphere of water supply, the National Environment Protection Action Program (1999) emphasises program measures; e.g., on the preparation of professional bases for the protection of sources of drinking water and the preparation of rehabilitation program for areas in which sources of drinking water are potentially endangered.

According to Čuček (2015), water from the public water supply is supplied to households and various economic activities (e.g., agriculture, mining, and quarrying, manufacturing, electricity supply) as well as for fire-fighting and road cleaning. Water for firefighting and road cleaning is not charged. Thus, the public service of the supply of drinking water is carried out by public companies based on authorities granted by local communities. These mandatory economic services are operated by using public infrastructure under their management. The procedure is described in the National Environment Protection Action Program (Nacionalni program varstva okolja, 1999).

Following the provisions of the Environmental Protection Act (ZVO-1, 2006) drinking water supply is required for municipal public service; therefore, 211 Slovenian municipalities are responsible for carrying out public service for water supply. Facilities and equipment necessary for the implementation of public water supply are part of the public infrastructure of local importance. The municipality is

Fig. 10.1 Drinking water supply in Slovenia (Statistični urad RS, 2015)

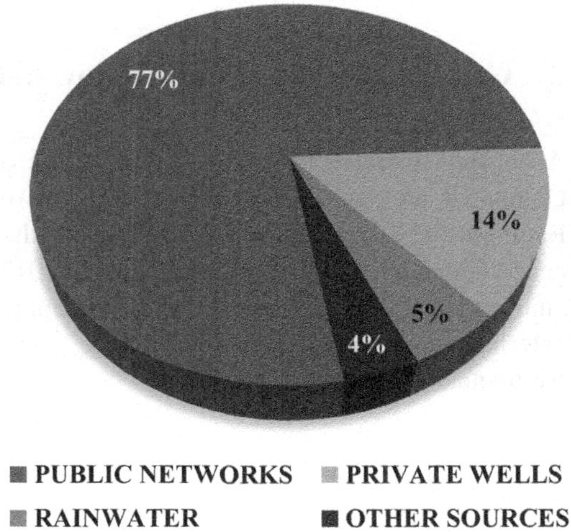

obliged to provide public service by the regulations governing utilities. Municipalities usually give the public appointment to utility companies (municipally owned) who provide water supply as public service. In 2013, 101 municipalities selected utility companies conducted a necessary public service municipal drinking water supply. List of those companies is available in the Information system of utilities public services for environmental protection. This register is established by Decree on drinking water supply (Uredba o oskrbi s pitno vodo, 2012).

In Slovenia, water is relatively cheap. According to Slaček (2016), costs for the use of water for drinking water supply in 2016 was approximately 6 cents per m^3 for beverage and technological purposes. For the needs of the bathing areas and natural spas, the cost was nine cents per m^3. Thus, the concession for thousands of litres of bottled groundwater for 2016 is 1.91 €, which is 6 cents lower than in the year 2015. The state expected that the remuneration for the needs of bathing, heating will receive little less than one and a half million Euros in the year 2017 and that beverage producer will pay more than 330,000 €. Nonetheless that the water is a natural public asset, there are cases of water privatisation in Slovenia. Due to the relatively low prices of water resources in Slovenia, the companies for the production of beverages are exciting to foreign corporations because of the access to water resources, allowed by the concession contract. Eman et al. (2016) stressed that such case occurred in 2015 when the Holland Company Heineken bought the Slovene brewery Pivovarna Laško. The primary reason was access to water resources. Later it was revealed that the company was interested only in the concession for the beverage production. Another similar case from Slovenia was when one foreign private company from Serbia bought company Fructal and the second foreign company from the Czech Republic bought company Radenska—two of the largest beverage Slovenian companies. The fear of public that the water sources would be claimed by multinational companies, which would begin to market the water as trade goods, was one of the reasons that the right to drinking water was adopted into the constitution.

2 Crime in the Republic of Slovenia

According to Baloh (2018), Slovenia is one of the safest countries in the world. Compared with other countries, it also has one of the lowest crime rates. The Republic of Slovenia was listed on one of the first places in the Global index of peace (Institute for Economics & Peace, 2017), which is measured based on the amount of organized internal conflict, violent protests, the level of crime, as well as external international relations of the country, military spending, and other parameters.

2.1 *The General Trend of Crime in the Republic of Slovenia*

Slovenia is mostly a 'transit country' for prohibited activities, especially for organised crime. Due to its geographical position (specific geographic location on the route from East to the West) and its history, Slovenia (also as a former Yugoslav republic) represents an important transit country. The so-called Balkan route runs across Slovenian territory; the route is directed from Eastern towards Western Europe and is used for the transportation of illicit drugs and stolen and looted works of art, as well as for the transportation and smuggling of migrants. Organised crime has been recorded since 1995 in Slovenia. The problem present narcotics trafficking, auto theft, fraud, tax evasion, counterfeit goods, alien smuggling but less than in neighbouring countries (OSAC, 2018). Meško et al. (2016) found out that the Slovenian statistics show that organised crime groups are mainly involved in drug-related crimes, followed by illegal trespassing of the state border and various types of grand larceny (though both are in the decrease). Organised crime groups are active primarily in major cities (e.g., Ljubljana and Maribor). Slovenian organised crime groups can generally be divided into two categories. The first includes groups that engage in traditional activities (e.g., drug trafficking, human smuggling); the second involves organised white-collar crime groups. In this latter case, economic crimes are committed by the top echelons of Slovenian society including a risky business of environmental deviance.

Crime in Slovenia is predominantly targeted at property rather than physical assault. Property is involved with around 87% of all criminal activities. According to the Police (Policija, 2018), there are, on average, 84,478 crime cases a year in Slovenia. The number of criminal offences from the year 2007 to 2017 are presented in Fig. 10.2.

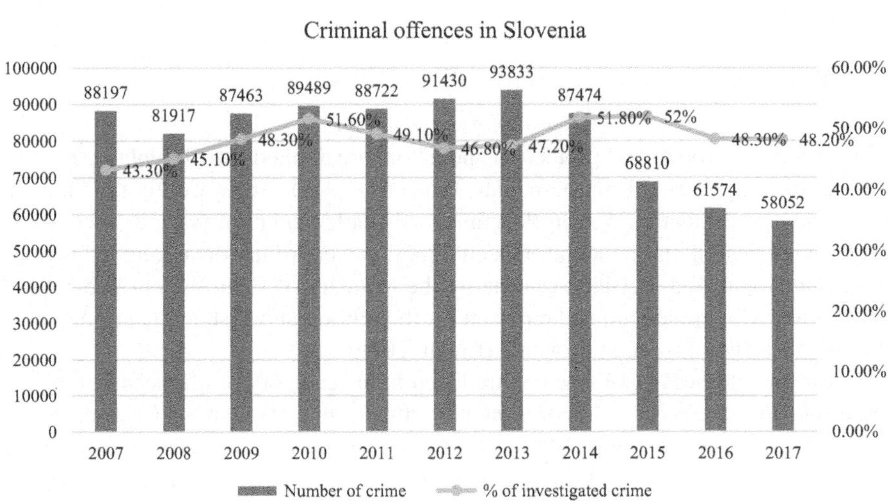

Fig. 10.2 Recorded criminal offences in Slovenia (Policija, 2018)

Figure 10.2 shows that the number of criminal offences fluctuated during the decade, while their number has dropped significantly in the last 2 years. The reason for the decline in the number of crimes in the last 2 years can be attributed to a police strike. In addition to the criminogenic factors, the effectiveness of the police and other authorities in detecting crime, influence the number of recorded crimes. The significant impact on the number of reported crimes also has the willingness of victims to report criminal offences.

According to Japelj (2017), the Slovenian police recorded 61,574 criminal offences in the year 2016, which is 10.5% less than in the year before. The proportion of criminal offenses that police have detected by its activity decreased from 15.2% in the year 2015 to 13.5% in the year 2016. The share of criminal offences investigated decreased from 52.0% to 48.3%. The police investigated 18.8% fewer economic crimes than in the year before. The scope of criminal offenses against sexual integrity increased 38.0% in the year 2016, while juvenile crime decreased by 3.0%. The Police dealt with 1.6% more criminal offenses against the economy than in the year before.

Japelj (2017) emphasise that the fight against economic crime is one of the priorities of the Slovenian police in last years. Especially in areas with a higher risk of economic and corruption crimes (banking, public authorities, health care, and protection of the EU financial interests). Police have primarily dealt with offences that have severely violated the rights of workers and vulnerable groups, while also raising awareness of their rights.

Transparency International's corruption perception index (COI) shows (Transparency International, n.d.) that Slovenia was ranked 35th of 168 countries in 2015 (up from 43rd in 2013). It achieved the rating of 60 in the range between 0 and 100 (being the result of "0" indicates a highly corrupt country, the result of "100" immaculate state).

3 Water Crimes in the Republic of Slovenia

The task of environmental protection and nature are defined in different documents: legislative (e.g., The Environmental Protection Act), software (e.g., National Environment Protection Action Program) and implementation (various measures). The role of design, implementation techniques, and coordination among the various documents with EU legislation borne by the Ministry of Environment and Spatial Planning. EU membership is the reason for the vibrant and systematic arrangement of environmental justice in Slovenia (Eman, 2016).

Criminal offences against water are listed in the category of offences against the environment. The scope of environmental crime (Table 10.1) in the Criminal Code of the Republic of Slovenia (2008) is defined in Chapter 32 as crimes against the environment, space and natural resources. The chapter covers 15 offences (Articles 332–347):

Table 10.1 Statistical data on water crime and crime against the environment and all criminal offences commited in the Republic of Slovenia from 2007 to 2017 (Policija, 2018)

	Year											
	2007	2008	2009	2010	2011	2012	2013	2014	2015	2016	2017	Total
Burdening and destruction of environment and space	9	14	57	31	33	21	47	34	42	25	30	343
Marine and water pollution by ships	–	–	0	0	1	0	–	–	1	–	n.d.	2
Contamination of drinking water	4	4	3	0	2	4	5	5	–	–	n.d.	27
Total	13	18	60	31	36	25	52	39	43	25	30	372
Criminal offences against the environment, space and natural goods	116	145	201	169	196	152	180	100	254	–	n.d.	1513
All criminal offences in the Republic of Slovenia	88.197	81.917	87.463	89.489	88.722	91.430	93.833	87.474	68.810	61.574	58.052	896.961

n.d. no data

- Burdening and Destruction of Environment (Article 332),
- Pollution of Sea or Waters from Ships (Article 333),
- Import and Export of Radioactive Substances (Article 334),
- Pollution of Drinking Water (Article 336),
- Tainting of Foodstuffs or Fodder (Article 337),
- Unlawful Occupation of Real Property (Article 338),
- Destruction of Plantations by a Noxious Agent (Article 339),
- Destroying of Forests (Article 340),
- Torture of Animals (Article 341),
- Game Poaching (Article 342),
- Fish Poaching (Article 343),
- Unlawful Handling of Protected Animals and Plants (Article 344),
- Transmission of Contagious Diseases in Animals or Plants (Article 345),
- Production of Injurious Medicines for the Treatment of Animals (Article 346)
- Unconscientious Veterinary Aid (Article 347).

According to statistical data, water crimes in Slovenia are not very common. Official statistics on crime are presented in Table 10.1.

Table 10.1 shows that the water crimes are not numerous in Slovenia and that the water crimes represent a small part (a quarter or even less) of all criminal offences against the environment. If we compare the number of all criminal offences reported in the Republic of Slovenia over the last 10 years with the number of reported water crimes, the percentage is minimal (0.04%). The contamination of drinking water and pollution of rivers, lakes, and ponds (included in the group 'Burdening and destruction of environment and space') is registered from five to ten times per year.

The analysis of Slovenian police reports on cases reported and investigated between 2005 and 2015 showed that the dominated water crime was water pollution (85%). Most often, the object of the offence is drinking water and free water in nature, such as rivers, lakes, and ponds. Potential dangers for the water sector are groundwater pollution, pollution of streams and rivers from factories, the theft of water from the pipeline system and the privatisation of water resources. In most cases, the perpetrators are unknown. The characteristics of the majority of suspects in water-related crimes in the police statistics were male (82%), females were suspects in 8% and in 10% of recorded water crimes a suspect is a legal person. The average age of suspects was 34–44 years old. The most significant problem presents a dark figure of crime against water because many cases are not detected or reported. As we already mentioned, some river sections are still burdened with excessive quantities of industrial and urban wastewater and are therefore contaminated or even over-polluted. Groundwater is contaminated with nitrates and pesticides, and additionally with chlorinated organic solvents at the local level (Ministrstvo za notranje zadeve, 2017).

According to Kozelj and Drev (2017), 91% of the population is supplied with drinking water through public water supply systems. The remaining 9% of the population is provided with water from their water catchment (wells and springs). The population supplying water from their water catchment must pay attention to the possible deliberate contamination of water, as it is relatively easy to commit and according to the data obtained from the police reports, is a widespread crime in Slovenia. This kind of crime occurs because of neighbourhood disputes or disagreements. The frequent reason for contamination of water resources is when the slurry flows into the water catchment area or the groundwater.

The data obtained from the Inspectorate for the Environment and Spatial Planning (Inšpektorat Republike Slovenije za okolje in prostor, 2016) stressed that on average, 675 different measures were issued on water, in the years 2013, 2014 and 2015 (warnings, payment orders, and offence decisions) (Fig. 10.3). The inspectors made control of the regulations of water protection area in most cases. They control applications for water permission. The most frequently identified violations in the field of public service treatment and disposal of urban wastewater are related to the program of care and record keeping. It was also recorded that there are still some municipalities, which do not guarantee the implementation of the mandatory public service. Their role is also to check water quality, and they are carried out controls imposed. They are regularly providing control of monitoring emission of substances and heat in the discharge of industrial wastewater into the public drains. The most significant problem in crime against water is the dark field, as many cases are not reported or perceived by law enforcement agencies. Given that we have reviewed the cases that took place between 2005 and 2015, it would be wise to extend future research to a more extended period.

Media often report about water pollution. This information is not surprising because they received information from the public. People inform the media at the moment when they notice some irregular patterns (e.g., different colour of water, the unusual smell of water, dead fish) on the river, pond, lake or other water recourses. The media receive the information also in cases when people found out that a person is stealing water from pipelines or persons get sick because of water

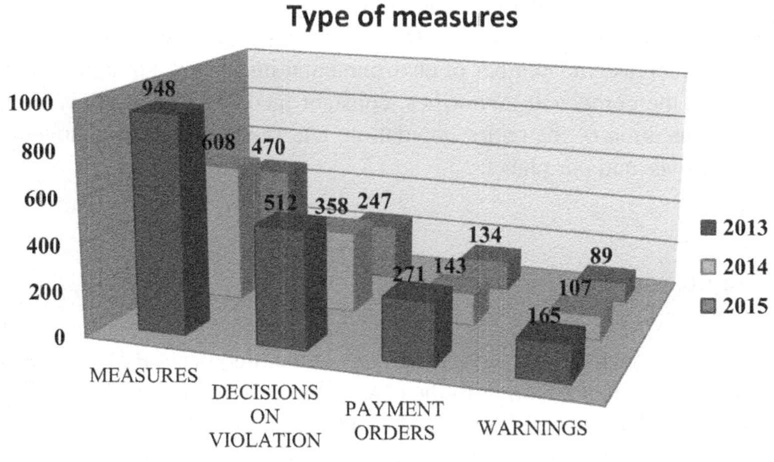

Fig. 10.3 Statistical data on the inspection measures from 2013 to 2015 (Inspectorate for the Environment and Spatial Planning, 2016)

pollution. There are also cases when journalists informed inspectors that an extraordinary event was committed (a case of at the location of small municipal wastewater cleaning device Vrzdenec in the year 2015) (Inšpektorat Republike Slovenije za okolje in prostor, 2016).

The analysis of the reports by the Inspectorate for the Environment and Spatial Planning and criminal statistics on crimes against the environment, space and natural resources and of the Slovenian police show, that the water crime (in most cases water pollution, intensive agriculture, industrial emissions, uncontrolled discharges of wastewater) is the prevalent environmental offence in Slovenia (Eman & Meško 2013; Eman, Meško, Dobovšek, & Sotlar 2013). The results reveal that water crime is not a frequent crime in the Republic of Slovenia compared to other countries. However, we believe that any such criminal offence or any pollution is too much, as it threatens the health and well-being of people, animals, plants and the environment in general. It should be emphasised that police and inspectorate data are only the data recorded by the repressive institutions. We must be aware that many water crimes remain unregistered, unreported or even unrecognised and the perpetrators are not prosecuted.

The Inspectorate for the Environment and Spatial Planning has regularly pointed out in its annual work reports the problem of the insufficient number of inspectors in the last years. Another issue is extensive and very contextual diverse legislation, which is continuously changing, and the associated enormous scope of work (Inšpektorat Republike Slovenije za okolje in prostor, 2016).

According to all data, we can see that the cases of water pollution predominate. However, these are the cases that were reported to the police and the inspectorate. The dark field of crime is enormous and will remain unknown. So far, in the Republic of Slovenia, only analyses have been carried out in the area of water crime. It would be recommended to consider some studies on water crime in the future that would give a more precise insight into the crime against water in the Republic of Slovenia.

In the future, active, immediate, and more stringent action should be taken by the competent authorities. Higher fines and stricter legislation should be adopted. It is necessary to increase the number of environmental inspectors, strengthen cooperation between the competent authorities, constant media coverage of water crimes and better education of the entire population, and raising awareness how water is vital for our lives and our planet.

4 Conclusion

Water is an essential asset for survival and one of the fundamental conditions of life. If we do not pay much attention to the care of the environment and water, the survival on the planet Earth will become extremely difficult. Brisman and South (2012) stressed that there are two extremes in the world of water. On the one hand, there is a lack of water, the consequences of which are reflected in poverty, starvation, and death. On the other hand, there are unpredictable large quantities of water (e.g., floods, tsunamis) that cause pollution of drinking water sources and consequently various diseases and death. The reason for such extremes is climate change caused by global warming and human interference in nature.

Eman and Meško (2012) found that the advancement and development of technology have reached the critical limits of the utilisation of the environment, which makes the person move along a thin border. It is necessary to realise that only a dash is needed and humankind will start sliding against the destruction of civilisations. Water is one of the first critical milestones. The consequences of crime against water can be catastrophic for all living creatures in the exposed area, and the devastating effects are visible for decades.

The geographical position of the Republic of Slovenia results in the fact that most of its territory is subject to relatively high annual precipitation is the reason that Slovenia is rich in water resources. Unfortunately, these resources are not evenly spatially distributed. The water governance in Slovenia is the responsibility of the state. However, citizens have an essential role in the process of implementation in the field of the environment. The public can also participate in the water management and water governance processes in the Republic of Slovenia. This possibility is significant because the public is commonly in a passive position when it comes to notification and communication. Its role in active inclusion is to participate in the shaping of content and influencing the final decision-making. The public also had an essential role in the process of entering the right to drinking water into the constitution. The Republic of Slovenia was the first country that adopted a human right to water into a constitution. It is essential to put the right to safe drinking water into the constitution because water is a vital part of the environment, a primary commodity and by its nature a given resource, which is of strategic importance to human life, development, prosperity and the security of the country. If the right to water is accepted as a human right, water will have to be accessible and enabled. Given the fact that the right to drinking water is adopted in the Constitution of the Republic of Slovenia, there is still a lot to be done in the area of water protection and management of groundwater.

The regulation in the Republic of Slovenia is quite extensive, which can pose some problems for the inspectors of the Inspectorate of the Republic of Slovenia for the Environment and Spatial Planning. The Police data on criminal offence charges shows that water crimes are not numerous in the Republic of Slovenia. The research of Slovenian police reports showed that water pollution predominates. Most often, the object of the offence is drinking water and free water in nature. In most cases, the perpetrators are unknown. If the suspect is known, male suspects prevail. The average age of suspects was 34–44 years old. Unfortunately, some rivers in the Republic of Slovenia are still burdened with excessive quantities of industrial and urban wastewater. The inspectors from the Inspectorate for the Environment and Spatial Planning recorded on average 675 offences per year. In most cases, they enforced the regulations of water protection areas and controlled applications for water permission. According to the literature review, research of the police reports and data from the inspectorate and articles published by the media, we can say that the dark field in the area of water crime is high.

Water protection should be an international, national and local priority, and the society has to carefully treat water resources to protect nature and assure sustainable development in the future. The United nation's social development goals include priorities in relation to the quality of life (no poverty), zero hunger, good health and well-being, quality education, gender equality, clean water and sanitation, affordable and clean energy, decent work and economic growth, industry, innovation and infrastructure, reduced inequalities, sustainable cities and communities, responsible consumption and production, climate action, life below water, life on land, peace, justice and strong institutions, and partnership for the goals (UNDP n.d.). Many of these proclaimed goals are also related to water protection and social action oriented to the future.

References

Agencija Republike Slovenije za okolje [Environmental Agency of the Republic of Slovenia; ARSO]. (2008). *Kakovost voda v Sloveniji* [Water quality in Slovenia]. Retrieved July 20, 2017, from http://www.arso.gov.si/vode/poro%C4%8Dila%20in%20publikacije/kakovost%20voda/Kakovost%20voda-SLO.pdf

Agencija Republike Slovenije za okolje [Environmental Agency of the Republic of Slovenia; ARSO]. (2015). *Količinsko stanje podzemnih voda v Sloveniji* [Quantity of groundwater in Slovenia]. Retrieved August 15, 2017, from http://www.arso.gov.si/vode/podzemne%20vode/publikacije%20in%20poro%C4%8Dila/Kolicinsko_stanje_podzemnih_voda_v_Sloveniji_OSNOVE_ZA_NUV_2015_2021.pdf

Agencija Republike Slovenije za okolje [Environmental Agency of the Republic of Slovenia; ARSO]. (n.d.). *Vode* [Water]. Retrieved August 10, 2017, from http://www.arso.gov.si/vode/

Baloh, M. (2018). *Slovenia: A country with one of the lowest crime rates in the world.* Retrieved August 28, 2018, from https://data.si/en/blog/slovenia-a-country-with-one-of-the-lowest-crime-rates-in-the-world/

Banjanac, L. S. (2013, February 14). Voda v Sloveniji je že "sprivatizirana" [Water in Slovenia is already "privatized"]. *MMC RTV SLO*. Retrieved June 20, 2017, from http://www.rtvslo.si/slovenija/voda-v-sloveniji-je-ze-sprivatizirana/302338

Brisman, A., & South, N. (2012). A green-cultural criminology: An exploratory outline. *Crime Media Culture, 9*(2), 115–135.

Brumen, J. (2017). Realisation of the principle of public participation in the water governance in the Republic of Slovenia: How to govern common water resources? In A. A. Lukšič (Ed.), *Exploration of political ecology in Slovenia* (pp. 52–69). Ljubljana: Fakulteta za družbene vede.

Čuček, S. (2015). *Gospodinjstva so v 2014 porabila za 0.1% več vode iz javnega vodovoda kot leto prej* [Households consumed 0.1% more water from public water supply than in the previous year]. Retrieved May 16, 2017, from http://www.stat.si/StatWeb/News/Index/5541

Direkcija Republike Slovenije za vode [Slovenian Water Agency]. (2016). *Slovenian water agency starts operating.* Retrieved September 10, 2016, from http://www.dv.gov.si/en/medijsko_sredisce/news/5645/

Direktiva Evropskega parlamenta in Sveta [Water Framework Directive]. (2000). *Uradni list EU*, (L 327).

Eman, K., Meško, G., Dobovšek, B. & Sotlar, A. (2013). Environmental crime and green criminology in South Eastern Europe - practice and research. *Crime, law and social change: an international journal. 59*(3), 341–358.

Eman, K. and Meško, G. (2013). Green issues in South-Eastern Europe. In N. South & A. Brisman (Eds.), *Routledge international handbook of green criminology* (pp.230–240). London;

New York: Routledge.Eman, K. (2016). Water crimes - a contemporary (security) issue. *Crimen, 7*(1), 44–57.

Eman, K., Kuhar, S., & Meško, G. (2016). Providing water supply in local communities – Case study of Slovenia. In G. Meško & B. Lobnikar (Eds.), *Criminal justice and security in Central and Eastern Europe. Safety, security, and social control on local communities* (pp. 367–374). Faculty of Criminal Justice and Security: Ljubljana.

Eman, K., & Meško, G. (2012). Ekološka kriminologija–veda o ekološki kriminaliteti [Ecological criminology-science on ecological crime]. In G. Meško, K. Eman, & A. Sotlar (Eds.), *Ekološka kriminaliteta in varovanje okolja-multidisciplinarne perspektive* (pp. 37–69). Ljubljana: Fakulteta za varnostne vede.

Eurostat. (2016). *Women in the EU earned on average 16% less than men in 2016.* Retrieved June 15, 2017, from http://ec.europa.eu/eurostat/en/web/products-press-releases/-/3-07032018-BP

Inšpektorat Republike Slovenije za okolje in prostor [Inspectorate for the Environment and Spatial Planning]. (2016). *Statistični podatki o inšpekcijskih ukrepih 2013 – 2015* [Statistical data on the inspection measures from 2013 till 2015]. Ljubljana: Inšpektorat Republike Slovenije za okolje in prostor.

Institute for Economics & Peace. (2017). *Global Peace Index.* Retrieved 20 August, 2018, from http://visionofhumanity.org/app/uploads/2017/06/GPI17-Report.pdf

Japelj, B. (2017). Kriminaliteta v Sloveniji leta 2016 [Crime in Slovenia in 2016]. *Revija za kriminalistiko in kriminologijo, 68*(2), 69–99.

Kazenski zakonik Republike Slovenije [Criminal Code of the Republic of Slovenia] [KZ-1]. (2008). *Uradni list RS*, (55/08, 66/08, 39/09, 55/09, 50/12, 54/15, 38/16, 27/17)

Kozelj, D., & Drev, D. (2017). Možnost kontaminacije pitne vode in varnostni ukrepi [Possibility of drinking water contamination and safety measures]. In L. Globevnik & A. Širca (Eds.), *Drugi slovenski kongres o vodah (Zbornik)* (pp. 141–146). Ljubljana: DVS – Društvo vodarjev Slovenije.

Meško, G., Sotlar, A., Dobovšek, B., Eman, K., Modic, M., Ažman, B., et al. (2016). Slovenia. In E. U. Savona & G. Berlusconi (Eds.), *Organized crime infiltration of legitimate businesses in Europe: A pilot project in five European Countries* (pp. 50–56). Trento: Transcrime.

Ministrstvo za notranje zadeve, Policija. (2017). *Policijska poročila o kriminalitete zoper vode med leti 2005 in 2015 [Police reports about water crimes in Slovenia from 2005 till 2015].* Ljubljana: Generalna policijska uprava.

Ministrstvo za okolje in prostor [Ministry of the Environment and Spatial Planning]. (1998). *Površinski vodotoki in vodna bilanca Slovenije (obdobje 1961 - 1990) [Surface streams and water balance in Slovenia].* Ljubljana: Hidrometeorološki Inštitut Slovenije [Hydrometeorological Institute of Slovenia].

Nacionalni laboratorij za okolje in hrano. (2016). *Letno poročilo o kakovosti pitne vode v letu 2014: Pitna voda v Sloveniji* [Annual report about the quality of drinking water in 2014: Drinking water in Slovenia]. Retrieved July 15, 2017, from http://www.mpv.si/uploads/PR14MonitPV_zakljucno_porocilo.pdf

Nacionalni program varstva okolja [National Environment Protection Action Programme]. (1999). *Uradni list RS*, (83/99, 41/04).

OECD. (2017). *Better Life Index: Slovenia.* Retrieved August 28, 2018, from http://www.oecdbetterlifeindex.org/countries/slovenia/

OSAC. (2018). *Slovenia 2018 Crime & Safety Report.* Retrieved July 9, 2019, from https://www.osac.gov/pages/ContentReportPDF.aspx?cid=23167

Policija. (2018). *Statistični podatki o kriminaliteti zoper okolje in vode 2007 – 2017* [Statistical data about crimes against the environment and water crimes in Slovenia from 2007 to 2017]. Ljubljana: Generalna policijska uprava.

Pravilnik o oskrbi s pitno vodo [Rules on drinking water supply]. (2006). *Uradni list RS*, (35/06, 41/08, 28/11, 88/12).

Pravilnik o pitni vodi [Rules on drinking water]. (2004). *Uradni list RS*, (19/04, 35/04, 26/06, 92/06, 25/09, 74/15, 51/17).

Pravilnik o zdravstveni ustreznosti pitne vode [Rules on drinking water quality]. (1997). *Uradni list RS*, (46/97, 52/97, 7/00, 19/04).

Slaček, N. (2016). *Tiha privatizacija vode* [Quiet water privatization]. Retrieved July 25, 2017, from http://radioprvi.rtvslo.si/2016/02/tiha-privatizacija-vode/

Slovenia.si. (n.d.). *Slovenia*. Retrieved July 20, 2018, from http://www.slovenia.si/slovenia/

Statistični urad Republike Slovenije [Statistical Office of the Republic of Slovenia]. (2015). *In 2014, households consumed 0.1% more water from the public water supply than a year earlier.* Retrieved August 28, 2018, from http://www.stat.si/StatWeb/en/News/Index/5541

Statistični urad Republike Slovenije [Statistical Office of the Republic of Slovenia]. (2018). *External trade in 2017 the largest in recent years*. Retrieved August 28, 2018, from http://www.stat.si/StatWeb/en/News/Index/7227

The Best Country Risk Report. (2017). *Slovenia*. Retrieved August 28, 2018, from http://www3.ambest.com/ratings/cr/reports/Slovenia.pdf

The European Environment Agency (EEA). (2015). *Slovenia country briefing - The European environment — state and outlook 2015*. Retrieved June 20, 2017, from https://www.eea.europa.eu/soer-2015/countries/slovenia

The Global Economy. (n.d.). *Slovenia: Political stability*. Retrieved July 10, 2018, from https://www.theglobaleconomy.com/Slovenia/wb_political_stability/

Transparency International. (n.d.). *Raziskave in indeksi* [Research and indexes]. Retrieved July 10, 2018, from http://www.transparency.si/dejavnosti/raziskave-in-indeksi

United Nations. (2017). *Sustainable development goals*. Retrieved July 10, 2018, from http://www.un.org/sustainabledevelopment/#

United Nations Development Programme [UNDP] (2015). *Human Development Report 2015*. Retrieved July 10, 2018, from http://hdr.undp.org/sites/default/files/2015_human_development_report.pdf

Uredba o kakovosti podzemne vode [Decree on the quality of underground water]. (2002). *Uradni list RS*, (11/02, 41/04).

Uredba o kemijskem stanju površinskih voda [Decree on the chemical status of surface waters]. (2002). *Uradni list RS*, (11/02, 41/04, 14/09).

Uredba o oskrbi s pitno vodo [Decree on drinking water supply]. (2012). *Uradni list RS*, (88/12).

Uredba o stanju podzemnih voda [Decree on groundwater status]. (2009). *Uradni list RS*, (25/09, 68/12, 66/16).

Uredba o stanju površinskih voda [Decree on surface water status]. (2009). *Uradni list RS*, (14/09, 98/10, 96/13, 24/16).

Ustava Republike Slovenije [Constitution of the Republic of Slovenia]. (1991). *Uradni list RS*, (33/91, 42/97, 66/00, 24/03, 69/04, 68/06, 47/13, 75/16).

Vlada Republike Slovenije [Government of the Republic of Slovenia]. (n.d.). *Osnovni geografski podatki* [Basic geographic data]. Retrieved June 20, 2017, from http://www.vlada.si/o_sloveniji/osnovni_geografski_podatki/

Water Information System for Europe (WISE). (2008). *Water Note 2: Cleaning up Europe's Waters*. Retrieved July 15, 2017, from http://ec.europa.eu/environment/water/participation/pdf/waternotes/water_note2_cleaning_up.pdf

Welch, C. (2018, March 5). Why cape town is running out of water, and who's next. *News.nationalgeographic.com*. Retrieved August 28, 2018, from https://news.nationalgeographic.com/2018/02/cape-town-running-out-of-water-drought-taps-shutoff-other-cities/

Zakon o ratifikaciji Konvencije o dostopu do informacij, udeležbi javnosti pri določanju in dostopu do pravnega varstva v okoljskih zadevah [Act Ratifying the Convention on Access to Information, Public Participation in Decision Making and Access to Justice in Environmental Matte]. (2004). *Uradni list RS*, (17/04).

Zakon o varstvu okolja [Environment Protection Act]. (2006). *Uradni list RS*, (39/06, 49/06, 70/08, 108/09, 48/12, 57/12, 92/13, 56/15,102/15, 30/16).

Zakon o vodah [Water Act]. (2002). *Uradni list RS*, (67/02, 2/04, 57/08, 57/12, 100/13, 40/14, 56/15).

Chapter 11
The Protection of Water from a Criminal Perspective: Water Crimes in Spain

Luz María Puente Aba and Eva María Souto García

1 Introduction

Studying water protection through criminal law requires the consideration of various perspectives to understand the numerous threats that may severely compromise the quality of water resources and their accessibility under equal conditions for the entire population. An area of water vulnerability concerns its sanitation, which is affected by pollution in general as well as by actions directly aimed at poisoning water intended for human consumption. Therefore, both *environmental crimes* and *crimes against public health* constitute criminal behaviour that poses a serious risk or even actual damage to the quality and safety of water as well as to human health and life. Using an economic perspective, we can also identify a second area of criminal actions that affect water as a consumer good: *theft or fraud* in water consumption alter access to this basic good under otherwise equal conditions, compromising its accessibility and availability.

Furthermore, it is worth noting how a third criminal category can become a factor that favours the commission of these different criminal offences that affect water: *corruption* by the public authorities, which can conceal or enable the commission of environmental, health or socio-economic crimes by legal persons. And lastly, it is interesting to note that the presence of water in human life means it can be turned into a medium through which certain particularly serious criminal behaviour can be committed: the case of *terrorism* has thus been identified, a criminal field including very diverse, objective-specific "modi operandi", one of which may be a damage or threat of the damage to waters intended for human consumption. Drawing from specific Spanish cases, this paper will analyse the profiles and characteristics of these five aforementioned criminal categories, all of which can seriously affect water intended for public consumption.

L. M. Puente Aba (✉) · E. M. Souto García (✉)
Faculty of Law, University of A Coruña, A Coruña, Spain
e-mail: lpuente@udc.es; esoutog@udc.es

K. Eman et al. (eds.), *Water, Governance, and Crime Issues*,
https://doi.org/10.1007/978-3-030-44798-4_11

2 Water Pollution: Water as a Target of Environmental Crime

When trying to identify one of the main and perhaps most obvious risks to water intended for human consumption, we must directly address the commission of crimes against the environment. Such crimes, broadly defined as part of the basic precept in the Spanish Criminal Code, involve the direct or indirect provocation or execution (in violation of the general environmental protection regulations) of "emissions, spillages, radiation, extractions or excavations, earthworks, noise, vibrations, injections or deposits into the atmosphere, the ground, the subsoil or the surface water, groundwater or seawater, including the high seas, even those affecting cross border spaces, as well as the water catchment basins which, by themselves or together with other incidents, cause or are likely to cause substantial damage to the quality of the air, soil or water, or to animals or plants", or "which may seriously damage the balance of the natural systems" (Article 325 of the Spanish Criminal Code).

Water is a fundamental element in the environment (Morillas Cueva, 2008; Silva Sánchez & Montaner Fernández, 2012; Suárez Mira Rodríguez, 2011), which can be seriously affected by criminal behaviour that constitutes pollution. As we will see, the vast majority of the judgement of criminal courts relating to environmental crimes refers to cases of pollution affecting water to a greater or lesser extent, even if the aim of the polluting conduct did not directly involve water.

Without going into the details on the specific environmental crimes included in the Spanish Criminal Code, the basic criminal behaviour in this area, as mentioned above, involves the commission of acts of pollution that "may seriously damage the balance of the natural systems"; this is the definition of so-called "ecological crime", established in Article 325 of the Spanish Criminal Code which contains an extensive list of different types of actions that could lead to environmental damage (Puente Aba, 2011). As can be seen, water is one of the fundamental elements of the environment, not only because all types of water are mentioned (surface, ground or seawater), but also because some forms of listed conduct imply by definition the involvement of water (notably, the reference to spillages, but also others such as deposits or earthworks) (Silva Sánchez & Montaner Fernández, 2012). When examining Spanish court rulings concerning environmental crimes, it can be noted that the vast majority of judgements refers to cases of discharge into surface water (rivers, lakes) or of leaks into groundwater (mainly due to deposits of polluting materials).[1]

[1]See the Spanish Supreme Court Judgement of 8th April 2008 (n° 141/2008), also, Spanish Supreme Court Judgement of 19th April 2010 (n° 289/2010), Spanish Supreme Court Judgement of 25th May 2016 (n° 453/2016) or, more recently, Spanish Supreme Court Judgement of 15th December 2016 (n° 941/2016).

Certainly, the gravest impact on water intended for human consumption results from the pollution of surface water and fresh groundwater (Górriz Royo, 2015).[2] However, the pollution of seawater can also have repercussions on the quality of water consumed by people. This fact can be analysed from two perspectives. On the one hand, it is important to bear in mind that seawater may be a source of drinking water intended for human consumption. Although this does not occur in all countries, in Spain at least, one of the ways of obtaining drinking water is through the desalination of seawater, which has also proven to be particularly useful in times of drought, like in 2017 (Ministry of Health and Social Care, 2009). On the other hand, the pollution of seawater can end up affecting water intended for human consumption, since the spread of pollutants in a medium such as water may harm other elements such as soil and surface water. As a paradigmatic example in Spain, we can cite the famous "*Prestige case*" (Martínez-Buján Pérez, 2003),[3] where the seawater of north-western Spain was contaminated with oil. Given a large amount of fuel discharged from the ship, the spill ended up affecting not only the Spanish coast but also the Portuguese and French coasts and possibly the surface and groundwater of those areas.

There are even some forms of pollution that indirectly affect water. The so-called "*Cercs case*",[4] the first case of environmental crime dealt with by the Spanish Supreme Court, is one such example. The case concerned a thermal power plant

---

[2] According to Art. 2 of the Council Directive 98/83/EC of 3 November 1998 on the quality of water intended for human consumption:

1. 'water intended for human consumption' shall mean:

(a) all water either in its original state or after treatment, intended for drinking, cooking, food preparation or other domestic purposes, regardless of its origin and whether it is supplied from a distribution network, from a tanker, or in bottles or containers;
(b) all water used in any food-production undertaking for the manufacture, processing, preservation or marketing of products or substances intended for human consumption unless the competent national authorities are satisfied that the quality of the water cannot affect the wholesomeness of the foodstuff in its finished form;

2. 'domestic distribution system' shall mean the pipework, fittings and appliances which are installed between the taps that are normally used for human consumption and the distribution network but only if they are not the responsibility of the water supplier, in its capacity as a water supplier, according to the relevant national law.

According to Art 2 of the Directive 2000/60/EC of the European Parliament and of the Council establishing a framework for the Community action in the field of water policy or, in short, the EU Water Framework Directive:

1. Surface water means inland waters, except groundwater; transitional waters and coastal waters, except in respect of chemical status for which it shall also include territorial waters.
2. Groundwater means all water which is below the Surface of the ground in the saturation zone and in direct contact with the ground or subsoil.

[3] See the Provincial Court Judgement of A Coruña of 13th November 2013 and the Spanish Supreme Court Judgement of 14th January 2016 (n° 865/2015).

[4] Spanish Supreme Court Judgement of 30th November 1990.

whose emissions were greatly exceeding the limits established by State regulations. The direct effect of these emissions was the serious pollution of around 30,000 hectares of forest land near the power plant. While it was proven that the emissions posed a serious threat to plant life in the area, it was not possible to prove that the emissions hurt the health of the people or cattle in the area or the waterways. However, it is reasonable to conclude that lakes or rivers within the affected hectares of land can be polluted as an effect of acid rain or seepage into the groundwater.

In almost all cases of these environmental crimes, the polluting act stems from the activity of a *legal person* (Puente Aba, 2011). This is logical given that the source of pollution usually comes from industrial, agricultural or livestock activities and the most commonly identified categories are listed with landmark cases for each one:

The dumping of waste from industrial activities: There are two important cases regarding this category. First, the Spanish Supreme Court Judgement of 1st February 2011 (n° 47/2011) convicted a scrap dealer for dumping scrap metal waste in a ravine, whose toxic agents subsequently polluted the river and groundwater. The second ruling, handed down by the Barcelona Provincial Court judgement of 19th February 2016 (n° 129/2016), convicted a company for discharging toxic solvents into the river Torelló and its surrounding aquifers.[5]

Leakage from landfills: This category is evidenced by the Spanish Supreme Court Judgement of 8th November 2011 (n° 1162/2011), which convicted those responsible from the municipal waste management company for causing the pollution of the river and the groundwater with their waste treatment practices.[6]

Leakage of organic waste from agricultural or livestock activities: The Spanish Supreme Court Judgement of 3rd April 2012 (n° 247/2012) concerned a livestock farm whose slurry discharges polluted the river and groundwater.

Pollution caused by aggregate extraction companies: One example case of this category is the Spanish Supreme Court Judgement of 18th October 2012 (n° 773/2012), which concerned a company dealing with the extraction and treatment of aggregates that caused pollution of the river and groundwater.[7]

These categories and rulings demonstrate that the source of the pollution lies in business activities and that the natural persons who had taken the decision to carry out or actually carried out the polluting act were convicted. However, it is important to bear in mind that the recognition of the criminal liability of legal persons, which has been present in the Spanish Criminal Code since 2010 (and which is also established in some other European legal systems), makes it possible to attribute criminal

[5] We can also cite the Spanish Supreme Court Judgement of 8th April 2008 (n° 141/2008), which convicted a company that sold dairy products; and which dumped its waste in a river near the factory.

[6] We can also cite the Spanish Supreme Court Judgement of 20th December 2007 (n° 1118/2007). In this case the company recycled lead batteries. See also the Spanish Supreme Court Judgement of 5th June 2009 (n° 600/2009).

[7] For a ruling on very similar events, see the Madrid Provincial Court Judgement of 30th March 2009 (n° 141/2009).

liability to both the specific natural person who commits the crime and also the legal person through which such criminal activity is carried out. Environmental offences are one of the criminal categories for which criminal liability may be imposed on legal persons; in fact, the first conviction of a legal person in Spain was for an environmental crime.[8]

The criminal liability of the legal person is independent of the liability of the natural person, and of course, they are not exclusive. Therefore, in the event of the commission of an environmental crime, as well as prosecuting the specific natural person(s) who committed the crime, the legal person must also be prosecuted, provided that the requisites for this form of liability are met.[9] What can certainly be complicated, bearing in mind that these are crimes committed within the framework of a corporate structure, is the identification of potential criminal liability on the part of individuals involved in the execution of the crime within the company's organisation chart.

Therefore, from the outset, it is not difficult to admit the criminal liability of the person who gives the order to carry out the polluting act (e.g., the person who orders the dumping of the waste into the river without first purifying it). Although the person who actually carries out this act may be a subordinate, (e.g., an employee of the company is ordered to deal with the waste management), it is clear that the criminal decision, the control over the act, lies with the person who holds the authority to decide and order the fate of the polluting materials. The very wording of the basic precept of environmental crime in the Spanish Criminal Code provides the basis for this interpretation, as it punishes anyone who "directly or indirectly causes or carries out" an act of pollution. We are dealing with an extensive concept of authorship that specifically contemplates the special singularity of criminal behaviour committed within the framework of the company, where the fundamental issue is the authority to make decisions on a certain matter, not the actual verification of the conduct (Górriz Royo, 2015; Martínez-Buján Pérez, 2008; Puente Aba, 2011; Quintero Olivares, 2013).

What is more, there has been some debate as to whether the person who carries out the order, i.e., the subordinate who follows the instructions of the superior and physically engages in action resulting in water pollution, can be held criminally liable. It has been argued that this subordinate carries out "neutral or standard behaviour", i.e., fungible behaviour that merely reflects the decision of a superior, who is the one who made the criminal decision.[10]

However, the business reality does not end in this simple division between manager and subordinate. Although the situation described above may arise in small legal entities (e.g., small agricultural or livestock farms), acts of pollution also come from large-scale companies (e.g., factories, power stations, etc.), which are made up

[8] Barcelona Provincial Court Judgement of 19th February 2014 (n° 155/2014), although this was a case of noise pollution.

[9] Articles 31 bis—31 quinquies in the Spanish Criminal Code.

[10] This is what happened, for example, in the case prosecuted in the Spanish Supreme Court judgement of 25th October 2002 (n° 1828/2002).

of top management and middle managers—both of whom are natural persons—granted the authority to make decisions on certain areas or sectors of the business activity (Faraldo Cabana, 2008).

In such a situation, in addition to holding the individual manager who took the criminal decision liable, it has been considered whether the managing director of the whole company, that is, the person who delegated the relevant authority, should also be held responsible. It can be argued that an intermediate position can be adopted according to which the delegating party may be held criminally liable under certain conditions. This requisite cannot be applied to all criminal acts carried out by individuals with delegated authority, since if this were the case, the phenomenon of the delegation of powers, which is vital for the operation of large companies, would be useless. In other words, it would be pointless to delegate a function to another individual, if the delegating party were to be held criminally liable for any decision taken by the individual delegated a function. In any event, it is acknowledged that the manager retains a "position of residual guarantee", which essentially implies two requirements; the manager delegates powers appropriately and maintains control over the delegate. With regards to the first requirement, appropriate delegation of powers implies that the manager has selected the delegate correctly, based on ability and trustworthiness to perform the function and the delegate must also be provided with the means and information necessary to carry out the work correctly. With regards to the second requirement, it is not enough to delegate appropriately and then simply disregard the delegate's activity; the manager must regularly monitor the delegate's activity to keep fundamental issues under control, such as maintaining the capacity and resources of the delegate, or approving decisions of vital importance for the company's activity. Thus, if there is a failure to comply with the requirements of this "position of residual guarantee", a manager may be considered the perpetrator or co-operator by omission with regards to the criminal conduct carried out by the delegate (Górriz Royo, 2015; Martínez-Buján Pérez, 2008; Puente Aba, 2011).[11]

A strikingly large proportion of convictions for environmental water crimes are for *negligent behaviour* (Górriz Royo, 2015; Ramón Ribas, 2011). In general, this is because the perpetrator does not directly intend to pollute water, but rather to minimise the costs of environmentally friendly business management, or at least to simplify the administration tasks of the company. As we have seen, pollution is usually caused by the uncontrolled release of certain substances or materials from the company's headquarters (e.g., industrial waste, slurry, fuel, etc.). In any event, we must bear in mind the fine line separating merely negligent behaviour from actions carried out "recklessly", i.e., from those cases in which the subject, although not having the direct aim of polluting, has no objective basis for excluding the possibility that his specific action will lead to water pollution (Hava García, 2002). It is thus important to analyse the specific spirit in which the offender acts, since the sanctions

[11] This is what happened, for example, in the cited Spanish Supreme Court Judgement of 25th October 2002 (n° 1828/2002), in the Spanish Supreme Court Judgement of 5th September 2001 (n° 1329/2001) or in the Spanish Supreme Court Judgement of 15th December 2016 (n° 941/2016).

that can be imposed vary greatly depending on whether the act is reckless or negligent.

3 Water Poisoning: Water as a Target of Crimes Against Public Health

Acts of pollution—a type of environmental crime—are not the only way through which the quality of water is worsened and made unfit for human consumption. In other situations, we may find ourselves dealing with specific cases of toxic agents being directly introduced into water intended for human consumption, with subsequent damage to people's health. In cases where there is no introduction of pollutants into the water, but rather the water is made toxic imminently before it is to be supplied to the population, we are faced with the possible commission of a *crime against public health*. This generic name covers a wide range of behaviours which, in essence, involve putting the health of the population at risk, mainly through the dissemination of toxic substances into goods intended for human consumption. The Spanish Criminal Code provides for a series of criminal acts within this scope, punishing various forms of tampering with food, medicines or beverages (including water) in such a way as to endanger people's health (Articles 359–367). Some specific examples related to the poisoning of water intended for human consumption can be cited.

Firstly, there was a case in Spain of bottled water being contaminated with organic wastewater, which caused an outbreak of gastroenteritis in a large number of people. The case did not reach court but was instead resolved through an investigation conducted by the health authorities, which determined the cause of the toxicity (the organic waste water) but failed to show how the bottled water was contaminated. The marketing company eventually compensated the injured parties, and the bottling company of this batch of bottles (from a spring in Andorra) closed its operations as a result of this event (El Confidencial, 2016; El Periódico, 2016). Secondly, there have been cases of the possible toxicity of tap water, the content of which may be altered by natural toxic agents and harm human health. Along these lines, for example, a case was reported of alleged contamination of tap water with natural arsenic, which could pose a risk to human health (Taboada, 2014).

These types of cases are usually due to the negligence of those responsible for water distribution, whether it is private companies or the public authorities, and serious negligence can be seen in the treatment and control of water intended for human consumption. However, although the original presence of toxins in water may be due to the negligence of the responsible individuals, the circumstances change when the individual becomes aware of the situation and fails to prevent it. From that moment on, the awareness and passivity of those responsible imply a clear will to carry out behaviour threatening to public health.

4 Water Theft or Fraud: Water as an Illicitly Gained Commodity

Criminal actions affecting access to water for human consumption may come from a perspective other than environmental damage or damage to health, namely an attack on the water as a commodity. This involves behaviour that does not affect the sanitation of water, but rather its accessibility and, in short, the fair distribution and balance of a limited good. There have been numerous cases of water intended for human consumption being stolen. In light of the specific characteristics of the distribution of water, such behaviour involves a certain technical complexity comprised of a surreptitious and fraudulent connection to the water distribution channels across any of its phases. Such acts may aim to obtain free tap water for domestic use (which may be accompanied by fraudulent access to electricity). However, it is more common to obtain larger quantities of water, which, therefore, "make up for" the risk and difficulties of carrying out this illegal access.

Spanish courts have tried cases of illegal procurement of water, either for domestic and private use (Almería Provincial Court Judgement, of 3rd September 2015, n° 345/2015) or jointly for domestic consumption and for agricultural and business use (Murcia Provincial Court Judgement, of 4th June 2015, n° 163/2015 or the Spanish Supreme Court Judgment of 11th November 2011 (n° 1169/2011). The latter form of illegal usage is evidenced by a case in which an agricultural farm supplied itself with water from a stream after it had illegally manipulated water channelling works. In this case, environmental damage was confirmed as a result of the destruction of the vegetation and natural elements near the stream.

Such water-threatening criminal behaviour will be more harmful when committed within the framework of a business activity, which will allow for legal persons to be held criminally liable. In addition to the issues already raised on this matter in Sect. 2, it bears mentioning that in the Spanish Criminal Code, legal persons can only be held criminally liable for some criminal offences. The Spanish Criminal Code establishes a system of "numerus clausus", choosing a closed set of offences for which a legal person may be held criminally liable. Although environmental crimes and crimes against public health are included in this list, the same is not true of the "electricity theft and the like"/"theft of public utility services" (Article 256 of the Criminal Code), a precept that would be applicable to the previously mentioned cases of illegal water procurement. It is, therefore, necessary either to revise the catalogue of offences for which legal persons may be held criminally liable, or to establish some "accessory consequence" for the company in cases where only the natural person can be convicted. For behaviour economically driven by the desire to significantly reduce costs, the action of the subject can only be intentional, and not an act of negligence.

While environmental crimes and crimes against public health violate supra-individual interests, which are fundamental for the social community, in economically driven behaviour, it is also possible to identify the involvement of collective interests. The improper procurement of water may be to the detriment of another legal person or private individual (e.g. another company or a neighbour), or the fraudulent mechanism may be implemented at a stage of the water supply before its arrival at a house or the site of a legal person. This will have implications for civil liability since the perpetrator of the offence will have to compensate either a specific person (who has paid for a certain amount of water that he has not consumed) or the State (who has supplied water to someone who has not paid for it). However, in both cases we can identify the impact on a collective interest, bearing in mind that water is a basic consumer good whose availability is limited, meaning that the removal of water could not only alter the possibilities of access to it, but it could also cause upward fluctuations in its price (especially in more extreme cases, such as during droughts).

However, even more complex situations may arise. During the long process of supplying water to the population, the public authority may outsource certain phases to private legal persons, which, as concessionaires of public service, will be in charge of managing the water supply to the citizens. Therefore, it is quite feasible that fraud involving illegal access to water is not simply an individual action, but may well involve a broader and more collective framework, even involving an agreement between consumers and the company supplying the service. Such fraudulent behaviour was reported in Spain, where a water managing company, in agreement with the end-users, charged them for the consumption of water for domestic use, when, in fact, the end-users utilised the water for industrial purposes, a use which should have been charged at a higher tariff (Agua de Cuenca, 2013; Martínez Cuartero, 2014) This case exemplifies possible fraud schemes that can be implemented within the scope of the supply of water for human use. Such criminal activities damage the State financially and cause evident deficiencies in water management.

5 Transversal Criminal Activity in Water Crimes: Corruption

Corruption, defined as the "abuse of power for private gain, both in the public and the private sector",[12] will also be present in certain actions of the public authority that refer both to the process of managing water intended for human consumption, and to the authorisation and control of private activities that pose a risk to water quality or to its accessibility.

Within the different categories of "water crimes" described so far, environmental crimes comprise the area with the broadest scope for verification. The Spanish Criminal Code (Article 329) outlines the specific offence of environmental

[12] See Comunicación de la Comisión al Parlamento Europeo, al Consejo y al Comité Económico y Social Europeo: Lucha contra la corrupción en la UE (COM(2011) 308 final, 6-6-2011) (p. 4, note 1), that takes this definition of corruption from the "United Nations Global Programme against Corruption".

malfeasance (abuse of environmental regulatory authority) punishing three types of conduct:

- Reporting favourably on the granting of unlawful permits that authorise the operation of industries or activities that carry out acts of pollution constituting environmental offences.
- Silencing the breach of environmental regulations when carrying out inspections in this area, or directly omitting the mandatory inspections.
- Ruling or voting in favour of the granting of unlawful permits.

The commission of environmental malfeasance is made possible by two factors. Firstly, certain public authorities are responsible for approving and monitoring new business activities which may produce potentially harmful pollution. Secondly, those public authorities are only supposed to approve the commission of new business activities if the company guarantees full compliance with all respective environmental standards. Thus, the offence of environmental malfeasance (abuse of environmental regulatory authority) punishes certain unlawful conduct carried out by authorities or public officers, who allow companies to pollute. Interestingly, Spanish legislation creates a very broad precept which: punishes administrative acts that involve the authorisation of the start-up of polluting companies (conduct that would already constitute the generic offence of prevarication); includes acts prior to the effective administrative authorisation, such as issuing favourable reports with a view to the future granting of such a permit; and even contemplates actions that only pose a risk of hypothetical damage to the environment, such as failing to carry out environmental inspections.

In any case, it is important to point out that this precept, although covering the sanctioning of a wide range of unlawful behaviour by public authorities or officers, may become ineffective if we take into account the evolution of administrative regulations on the authorisation and control of companies liable to cause pollution. Furthermore, the impact of the adaptation of the various State regulations on European Union law may change the premises on which the definition of this criminal precept is based. As we have seen, this form of environmental malfeasance (abuse of environmental regulatory authority) is based on certain types of administrative activities, such as the granting of authorisations and permits, and the issuing of reports before granting permits or the carrying out of inspections. In this respect, it is interesting to note the progressive loss of centrality of the requirement of prior authorisation or permit for the operation of a business activity, which in many cases is no longer necessary and has been replaced by the simple need for the employer to issue prior notification or a responsible declaration to the authority. Since the passing of Directive 2006/123/EC of the European Parliament and of the Council of 12th December 2006, this new system of prior notifications and responsible declarations has been consolidated on services in the internal market, which implies a change of model by replacing the need to apply for and obtain an administrative permit for certain activities with the mere need for the person concerned to issue notification or a declaration of the activity. This change of system, therefore, poses challenges when it comes to acknowledging a crime of environmental malfeasance, since in these cases the Authority does not have to decide or vote in favour of granting a permit. Instead, the mere submission of the prior notification constitutes an enabling document for the act to be carried out (modification or use of the land), and the Authority is simply empowered to declare the submitted documentation as complete or to request the correction of any possible deficiencies. Attention must be paid to the necessary harmonisation between the definition of criminal offences and the administrative regulations on this matter, so that failure by the authorities to comply with their controlling obligations cannot escape criminal sanctions.

The vast majority of these criminal behaviours that constitute corruption are based on collusion between the Authority and the company in such a way as to authorise or tolerate the production of polluting waste by the company. A clear example of this is the aforementioned Spanish Supreme Court Judgement of 8th November 2011 (n° 1162/2011). Along these lines, another notable case involves an alleged agreement between the administrative authorities and the management of a prison facility (in this case, all public officials) to discharge faecal wastewater from the prison into the nearby river without any purification (Europa Press, 2014; Huertas, 2013).

However, it would be feasible to verify negligent behaviour on the part of the Authority, which, in the event of failure to perform its duties, carries out flawed management resulting in the contamination of natural elements. Thus, for example, the alleged lack of maintenance of the former Bolidén mine in Aznalcóllar by the Authority (which, incidentally, caused one of the most significant environmental disasters in Spain in April 1998, which affected the Doñana natural park) was reported. The polluting impact of the toxic flood was estimated to be one hundred times greater than the 63,000 tons of fuel oil spilt by the Prestige—the environmental disaster off the Galician coast in the north of the Iberian Peninsula. Altogether, the incident wiped out 4386 hectares along 62 km of the Guadiamar river, the main tributary of the Guadalquivir river (El País, 2010; El Periódico, 2015), and caused toxic mineral residue from the mine to seep into the aquifers. A further example is the pollution of Lake Sanabria by the discharge of faecal wastewater, apparently caused by poorly maintained sewerage and wastewater treatment networks (20 Minutos, 2013; Sevillano, 2014).

In these cases, the responsible authorities or public officials are not the perpetrators of a crime of prevarication, since there is no negligent administrative prevarication in the Spanish Criminal Code. Moreover, many cases demonstrate no specific action constituting the issuing of a ruling or a report, but rather a deficiency in the management of certain activities or facilities. In any event, it is clear that the responsible authorities could have committed a crime against the environment, considering that the actions of the public authorities or officials cause pollution, especially bearing in mind that the Spanish Criminal Code sets out crimes against the environment broadly, with the basic precept punishing anyone found to cause or provoke, directly or indirectly, an act of pollution.

If we focus on fraudulent actions involving access to water, it is worth recalling the case described in Sect. 4 regarding agreements between end-users who consume water and the company providing the public service. In this type of case, we can verify corrupt behaviour either because the (private) company specifically responsible for providing the public service could be considered a genuine "public official", or because the agreement to defraud could be established directly with the public authorities or officials responsible for controlling water management and supply.

Lastly, given that corrupt behaviour can occur in any area of public authority, such as in an area dedicated to the management of public water services at any stage (procurement and storage, purification, supply, invoicing and wastewater), cases of corrupt activities with no effect on the quality or the invoicing of water may still negatively impact the adequate provision of water. An example of this is a major case of corruption in Spain that is still under investigation as of the writing of this article, the so-called *"Operation Lezo"*. The case involves alleged corrupt behaviour within the *Canal de Isabel II* public entity, which is responsible for supplying water in the Community of Madrid. The investigation focusses on the alleged diversion of public funds from the entity to Madrid's political posts, and whether certain acquisitions of Latin American companies procured by *Canal de Isabel II* were made at costs exceeding the market prices (El Confidencial, 2017; El País, 2018). If these allegations are corroborated, it constitutes corrupt behaviour that is not aimed at diminishing water quality or fraudulently obtaining water, but which constitutes the withdrawal of public funds intended to manage the complete water cycle, a crime which could hurt the adequate provision of this public service. In conclusion, the presence of corrupt behaviour in public authority sectors, or private companies that collaborate with the authorities in the provision of water services, may hurt the quality and accessibility of this essential resource for human consumption, even though the direct aim is not to affect water.

6 Water Terrorism: Water as a Means for Serious Attacks on the Population: The Case of Terrorism

In the previous cases, we described and analysed criminal behaviour that directly targeted water, either as a natural element that can be polluted or poisoned or as a consumer good that can be obtained fraudulently. However, given the importance of water in human life, it may not always be the ultimate focus or aim of criminal behaviour, but rather a means of committing certain crimes. This possibility can be confirmed by examining the "modus operandi" of one of the most alarming criminal behaviours in today's society—terrorism. Without going into the varying types or concepts of terrorism, we can start by looking at the recent Directive (EU) 2017/541 of the European Parliament and of the Council of 15 March 2017 on combating terrorism (replacing Council Framework Decision 2002/475/JHA and amending Council Decision 2005/671/JHA). In it, Article 3 defines terrorist offences and lists a number of criminal acts in paragraph 1 as terrorist offences if they are carried out for any of the purposes set out in paragraph 2: (a) seriously intimidating a population;

(b) unduly compelling a government or an international organisation to perform or abstain from performing any act; (c) seriously destabilising or destroying the fundamental political, constitutional, economic or social structures of a country or an international organisation.

Among the possible criminal acts mentioned in the paragraph above 1 are attacks upon a person's life which may cause death; attacks upon the physical integrity of a person; or release of dangerous substances, the effect of which is to endanger human life. Interestingly, the Directive explicitly mentions "interfering with or disrupting the supply of water, power or any other fundamental natural resource, the effect of which is to endanger human life" as a possible means of terrorism.

Therefore, terrorist action aimed at such objectives may take the form of poisoning water intended for consumption by the population. A possible case of this kind was detected in Spain in 2011, when an individual was arrested for attempting to poison the water in storage tanks intended to supply homes, hotels and campsites. The alarm was raised when his statements were found on social networks and Internet forums in which he defended the terrorist struggle, and revealed his intention to poison water intended for human consumption in certain areas of southern Spain. He was eventually convicted of the crime of extolling terrorism related to his statements made in defence of acts of terrorism, but no evidence was found that the threat to poison water would be carried out.[13] In any event, aside from the vicissitudes of this specific case, the use of drinking water as a means to harm people can be considered; in fact, we are once again faced with the above-referenced cases of water poisoning, analysed in Sect. 3.

By definition, terrorist acts involve clear and direct intent by the perpetrator. Such acts constitute a special case since, contrary to the other categories described in this paper, the economic motive (profit making, cost saving) is absent in these criminal behaviours, thus representing a new threat to the security and sanitation of water.

7 Conclusion

This paper has shown the different perspectives from which water can be affected by the commission of crimes. The most striking one is pollution and water poisoning, as they can make water unsuitable for human consumption and pose a risk or potential harm to human health. Such behaviour could constitute, in the Spanish Criminal Code, either a crime against the environment or a crime against public health. In this way, the law provides for the punishment of these acts and complies with the constitutional provisions for human life and environmental protection. Yet, an effective deterrence would be achieved if two specific issues were taken into

[13] National Court Judgement of 12th July 2013 (n° 24/2013).

account: the frequent cases of negligence and the involvement of public authorities in the commission of these crimes.

Many cases of water pollution or poisoning are due to negligent behaviour in enterprise management. Corrupt practices have been revealed as a factor that facilitates crimes affecting water. This corruption takes the form of either collusion with the perpetrator (usually an enterprise) in environmental crimes, or of facilitating the commission of these crimes and allowing for their impunity. Therefore, preventive measures for this kind of criminal offences would be, in the first place, the implementation of periodic inspections to monitor the enterprises environmental commitments, and in the second place, the prevention and prosecution of corruption in Public Administration.

Water accessibility and availability can not only be affected by the commission of crimes from the perspective of the environment and human health, but also from an economic one. For example, fraudulent behaviour and theft during the process of water distribution impede universal access to water, and are crimes punishable according to Spanish Criminal Code.

Finally, it is important to detect how water can be used as a means to commit other types of crimes that can affect human life. One additional type of crime is terrorism, whereby water is manipulated to pose a terrorist threat, regardless of whether it is real or fictitious. Such a crime could be punished using the comprehensive regulations of terrorism in the Spanish Criminal Code. This shows, once again, the need for implementing adequate and effective preventive measures to deter crime and enforce existing criminal regulations.

References

Minutos. (2013, December 13). El Lago de Sanabria pierde un 87% de biodiversidad por los vertidos contaminantes. *20Minutos.es*. Retrieved November 15, 2019, from https://www.20minutos.es/noticia/2004490/0/lago-sanabria/zamora/contaminacion/

Agua de Cuenca. (2013, December 12). *El fraude del agua*. Retrieved November 15, 2019, from https://aguadecuenca.wordpress.com/2013/12/12/el-gran-fraude-del-agua-en-cuenca/

El Confidencial. (2016, April 25). Restos fecales humanos provocaron el brote de gastroenteritis en 4136 personas. *El Confidencial*. Retrieved November 15, 2019, from https://www.elconfidencial.com/espana/cataluna/2016-04-25/agua-embotellada-con-restos-fecales-humanos-contaminacion_1189748/

El Confidencial. (2017, June 6). Caso Lezo: el "botín" de Ignacio González en Colombia y todo lo que se sabe hasta ahora. *El Confidencial*. Retrieved November 15, 2019, from https://www.elconfidencial.com/espana/2017-06-06/operacion-lezo-cronologia-noticias-ignacio-gonzalez_1374454/

El País. (2010, October 6). El desastre que amenazó Doñana. *El País*. Retrieved November 15, 2019, from https://elpais.com/diario/2010/10/06/sociedad/1286316006_850215.html

El País. (2018, June 20). La Comunidad de Madrid denuncia nuevos delitos en la gestión del Canal de Isabel II. *El País*. Retrieved November 15, 2019, from https://elpais.com/ccaa/2018/06/20/madrid/1529489567_403316.html

El Periódico. (2015, May 13). Aznalcóllar, una mina marcada por el desastre ecológico de Boliden de 1998. *El Periódico*. Retrieved November 15, 2019, from https://www.elperiodico.com/es/sociedad/20150513/aznalcollar-una-mina-marcada-por-el-desastre-ecologico-de-boliden-de-1998-4182436

El Periódico. (2016, April 25). Restos fecales humanos causaron el brote de gastroenteritis por agua envasada Eden. *El Periódico*. Retrieved November 15, 2019, from https://www.elperiodico.com/es/sociedad/20160425/agua-eden-intoxicacion-restos-fecales-5083365

Europa Press. (2014, December 6). Juez ve indicios de delito en vertidos de aguas fecales procedentes de la cárcel de Albolote. *Europa Press.es*. Retrieved November 15, 2019, from https://www.europapress.es/andalucia/noticia-juez-ve-indicios-delito-vertidos-aguas-fecales-procedentes-carcel-albolote-20141114142953.html

Faraldo Cabana, P. (2008). Problemas de atribución de la responsabilidad penal derivados de la estructura jerárquica de la empresa. In G. Quintero Olivares & F. Morales Prats (Eds.), *Estudios de Derecho Ambiental. Libro Homenaje al Profesor Miquel Prats Canut* (pp. 469–523). Valencia, Tirant lo Blanch.

Górriz Royo, E. M. (2015). *Delitos contra los recursos naturales y el medio ambiente*. Valencia: Tirant lo Blanch.

Hava García, E. (2002). *La imprudencia consciente*. Granada: Dykinson.

Huertas, Y. (2013, March 13). Denuncian un nuevo vertido de aguas residuales de la prisión de Albolote. *Granada Hoy*. Retrieved November 15, 2019, from http://www.granadahoy.com/granada/Denuncian-vertido-residuales-prision-Albolote_0_1007599655.html

Martínez Cuartero, A. (2014, January 28). El gran fraude del agua en Cuenca. *Periodistas en español.com*. Retrieved November 15, 2019, from https://periodistas-es.com/el-gran-fraude-del-agua-en-cuenca-28066

Martínez-Buján Pérez, C. (2003). Las posibles responsabilidades penales en el "caso Prestige". *Ambiente y Derecho, 1*, 11–32.

Martínez-Buján Pérez, C. (2008). Protección penal del medio ambiente y personas jurídicas. In G. Quintero Olivares & F. Morales Prats (Eds.), *Estudios de Derecho Ambiental. Libro Homenaje al Profesor Miquel Prats Canut* (pp. 577–620). Valencia: Tirant lo Blanch.

Ministry of Health and Social Care. (2009). *Guía de desalación: aspectos técnicos y sanitarios en la producción de agua de consumo humano*. Madrid. Retrieved November 15, 2019, from https://www.mscbs.gob.es/profesionales/saludPublica/docs/Guia_desalacion.pdf

Morillas Cueva, L. (2008). El agua como objeto de protección penal en su vertiente ambiental. In G. Quintero Olivares & F. Morales Prats (Eds.), *Estudios de Derecho Ambiental. Libro Homenaje al Profesor Miquel Prats Canut* (pp. 1067–1090). Valencia: Tirant lo Blanch.

Puente Aba, L. M. (2011). Los delitos contra los recursos naturales y el medio ambiente. In P. Faraldo Cabana (Ed.), *Ordenación del Territorio, Patrimonio Histórico y Medio Ambiente en el Código Penal y la Legislación Especial* (pp. 235–279). Valencia: Tirant lo Blanch.

Quintero Olivares, G. (2013). *Derecho Penal Ambiental*. Valencia: Tirant lo Blanch.

Ramón Ribas, E. (2011). Los delitos contra los recursos naturales y el medio ambiente. In P. Faraldo Cabana (Ed.), *Ordenación del Territorio, Patrimonio Histórico y Medio Ambiente en el Código Penal y la Legislación Especial* (pp. 358–368). Valencia: Tirant lo Blanch.

Sevillano, E.G. (2014, April 22). Vertidos en el Lago de Sanabria. *El País*. Retrieved November 15, 2019, from https://elpais.com/sociedad/2014/04/22/actualidad/1398195617_792557.html

Silva Sánchez, J. M., & Montaner Fernández, R. (2012). *Los delitos contra el medio ambiente*. Barcelona: Atelier.

Suárez Mira Rodríguez, C. (2011). La protección del Medio Ambiente en la Constitución Española. In P. Faraldo Cabana (Ed.), *Ordenación del Territorio, Patrimonio Histórico y Medio Ambiente en el Código Penal y la Legislación Especial* (pp. 55–76). Valencia: Tirant lo Blanch.

Taboada, P. (2014, March 28). El Seprona acusa a Aquagest de ocultar contaminación de agua en Ourense. *El País*. Retrieved November 15, 2019, from https://elpais.com/ccaa/2014/03/27/galicia/1395948205_343730.html

Chapter 12
Water Pollution and Contamination from Gold Mines: Acid Mine Drainage in Gauteng Province, South Africa

Anthony Minnaar

1 Introduction

In mines where, in particular, metal ores such as gold, silver and copper are mined, the accompanying sulphide minerals in the surrounding rock are exposed to water and air. This process, termed pyrite oxidation, results in ferrous sulphates (including iron) dissolving and is accompanied by the generation of sulphuric acid.[1] This acid then creates a further reaction in the surrounding rock by dissolving other harmful toxic metals and metalloids such as cadmium, arsenic, lead and mercury. So, actual mining increases the exposed surface area of sulphur-bearing rocks allowing for the excess acid generation beyond normal natural underground oxidation. Mine tailings and waste rock, having a much greater exposed surface area than ore underground, are more prone to acid formation since large masses of sulphide minerals are exposed quickly during the mining and milling processes (Jennings, Neuman & Blicker, 2008, pp. 3–4; Department of Water Affairs (DWA), 2017).[2]

When underground water seepage or heavy rainfalls occur, this harmful acid and its by-products are leached out of the underground mine tunnels or drained from waste rock piles/dumps, mine tailings, open pits and mine drainage dams or leach pads, thus entering both into underground water systems and into above-ground streams and rivers. This form of water pollution is termed Acid Mine Drainage (AMD). As early as 1987, the US Environmental Protection Agency had recognised that:

[1] The same caustic substance as used in car batteries.

[2] Note that in July 2014 this Department was renamed as the Department of Water and Sanitation having taken over the sanitation function from the Department of Human Settlements.

A. Minnaar (✉)
Department of Criminology and Criminal Justice, University of Limpopo, Sovenga, South Africa

K. Eman et al. (eds.), *Water, Governance, and Crime Issues*,
https://doi.org/10.1007/978-3-030-44798-4_12

> …problems related to mining waste may be rated as second only to global warming and stratospheric ozone depletion in terms of ecological risk. The release to the environment of mining waste can result in profound, generally irreversible destruction of ecosystems (US Environmental Protection Agency (EPA), as cited in Council for Scientific and Industrial Research (CSIR), 2009, p. 1).

Accordingly, the resulting acid mine drainage under such mining conditions worldwide is one of the most serious threats to water systems and can lead to toxic contamination of rivers, streams and aquatic life for many years. But it also has a serious impact on agriculture and farming livelihoods, as well as being a threat to the health of people and animals. Acid mine drainage is especially harmful since it can continue indefinitely. In other words, AMD can cause environmental damage and threaten the health and safety of nearby communities long after a mine closure occurs. This pollution is so persistent that, in the absence of available remedies, in many instances, the contaminated sites may never be completely restored (Ochieng, Seanego, & Nkwonta, 2010, pp. 3351–3352).

The toxic metals contained in AMD water are particularly problematic since they do not break down in the environment. In addition, they tend to settle on river bottoms and stream beds. As a result, they have a long-term lingering presence in water systems. This ensures a lasting source of contamination to the aquatic insects living in river systems and to the fish that feed on them (Earthworks, 2017, p. 1). Water contaminated by Acid Mine Drainage: "…can be toxic …leaving receiving streams devoid of most living creatures. [AMD] …is responsible for physical, chemical, and biological degradation of stream habitat" (Jennings, Neuman & Blicker, 2008, p. 5).

There are also several collateral consequences to AMD toxicity in ground water and river systems, namely: (1) contamination of the food chain with the accumulation of heavy metals in the tissues of fish and other creatures. If enough metals collect, it can be toxic to the fish or to any creatures (or humans for that matter) that eat such contaminated fish; (2) pollution of drinking water supplies due to the toxic heavy metals that remain dissolved in the acidic water from mines. These metals can be absorbed by humans drinking such contaminated water. This can cause severe health problems; (3) deterioration of ecosystems. This can occur as damage to wildlife and water systems in drier regions, impacting on all aspects of the ecosystem since arid regions have very delicate and fragile balances, with very little leeway absorbing any increased levels of acidity in the water supply due to AMD (United States Geological Survey, 2000).

In several of the major mining countries in the developed world there exists legislation that requires such acid mine drainage waters to be treated (neutralised) before being discharged into streams and river systems. For example, the US,[3]

[3] In the US, The Clean Water Act (CWA) (1972) establishes the basic structure for regulating the discharges of pollutants into the waters of the United States and regulating quality standards for surface waters. The CWA also made it unlawful to discharge any pollutant from a point source into surface water systems, unless a permit was obtained. The National Pollutant Discharge Elimination System (NPDES), administered by the EPA, implemented pollution control programs and developed national water quality criteria recommendations for pollutants in surface waters. The NPDES

Canada[4] and Australia[5] have, over the years, developed comprehensive guidelines, policies and regulations, linked to specific pieces of legislation, that deal with the treatment of AMD from mines. But this does not always occur, or if it does the resulting residue remains in so-called mine sludge dams (also known as slimes ponds). The danger here being that often, after such mines are closed or abandoned, the above ground sludge dams deteriorate over time and their retaining walls break open or overflow after heavy rains with the harmful sludge residues then again washing into water systems. The neutralisation process, largely using lime (which in itself is expensive), is not only a costly process, it also has several accompanying disadvantages, namely:

- The scaling of equipment by the unstable water produced; and
- Malfunctioning of lime-dosing equipment, i.e. the lime particles settle and line the inside of pipelines and valves which then often causes blockages. This can lead to either under-dosage of the water being treated, which then also leads to acid corrosion of equipment and pipes (Foundation for Water Research, 2004).

Such degradation of neutralising plant equipment on the mines reduces the neutralising plant's working life span requiring costly replacement. Accordingly, acid mine water treatment must continue for many years (in some cases in perpetuity[6]) and can be a significant economic burden to a country if a mining company files for bankruptcy or refuses to cover the cost of treating the mine water. Even with existing technology, acid mine drainage is virtually impossible to stop once the various reactions begin (Earthworks, 2017, pp. 1–2).

---

issues permits to all wastewater dischargers and treatment facilities. These permits establish specific discharge limits, monitoring and reporting requirements and may also require these facilities to undertake special measures to protect the environment from harmful pollutants, for example: AMD (United States Environmental Protection Agency (EPA), 2019).

[4] A Canadian government program called Mine Environmental Neutral Drainage (MEND) sited in the Environment Protection Agency of Environment Canada has been working since the late 1980s on reducing the risk and liability from acid mine drainage and checking that Canadian mining operations comply to the mining and clean water regulations in this regard (Jennings, Neuman & Blicker, 2008, p. 7).

[5] In Australia, the federal government and the state and territory governments have responsibilities to protect human health and the environment from the harmful effects of mining, including by mitigating potential adverse impacts from AMD. Accordingly, all Australian mining jurisdictions have enacted legislation for the assessment and management of mining—all in line with international environmental law. As a consequence, government agencies in each state or territory have a responsibility to ensure that mineral exploration, mining and mine closure take place in accordance with the relevant environmental protection legislation. The principal piece of Australian Government legislation that is used in this regard is the *Environment Protection and Biodiversity Conservation Act* 1999 (Australian Government, 2016, pp. 169–170).

[6] There are Roman mining sites in Great Britain that continue to leach out AMD water more than 2,000 years later (Earthworks, 2017, p. 1).

2 Gold Mining in Gauteng Province, South Africa[7]

In 1886 gold was discovered in South Africa. Since then gold mining has been a mainstay of the mining economy in that country. By 1970, South African gold production reached its high point, producing 65% of world production. But since then gold production has declined and by 2006, although at the time still the biggest single world gold producer, this had dropped to 12% of total world production. In 2007 South Africa dropped to 2nd position and 4th in 2009. By 2010 South Africa's position had dropped to 5th on the world list of gold producers (a position held until 2013) but in 2010 was still producing approximately 11% of the total world production—this had dropped to 4.2% in 2017, while between 2014 and 2018 South Africa had dropped to number 7th in world gold production. By the end of 2017 the Witwatersrand Basin still produced 93% of South Africa's gold output (U.S. Geological Survey, 2019; Minerals Council South Africa, 2018, p. 34; South African Government, 2019).

One of the consequences of the decline in gold mining production in South Africa has been the abandoning of unprofitable gold mines, which started in the 1970s but quickened up after the economic downturn in the global economy from 2008 onwards. The 2008 global financial and economic constraints made the closure of marginal mines occur earlier than planned which in turn resulted in the pumping of water out of mines to either cease or be scaled down to cut costs. But as pumps were shut down, it also hastened the flooding of the underground mine voids making further outflows of AMD unavoidable.[8]

The South African legislation on acid mine drainage requires that the sludge (high solids heavy metals content residue) from the neutralisation plants be discharged into lined dams or ponds (also a costly requirement in the neutralisation process) so as to prevent the harmful metal leachates from polluting underground water (Foundation for Water Research, 2004).

Due to these and other factors, several mining companies do not treat mine water or after a time abandon such water neutralisation operations due to the high costs involved. Furthermore, in the Gauteng Province of South Africa, over the last ten years (post-2008 global economic downturn) several mining companies in the gold mining sector have simply gone out of business. This was largely due to rising costs in mining operations overall but was compounded by the falling price of gold. As a consequence of these factors, being bankrupt or simply abandoning unproductive mines, these mining companies did not fulfil their legal obligations concerning the

[7] In a later section specific reference to the following two mines in the Eastern Basin of the Witwatersrand will be mentioned concerning the pumping of AMD water from their mines: Rand Uranium Mine; and the Pamodzi/Aurora Gold Mine (formerly Grootvlei Mine)—both on the East Rand.

[8] In fact, in late August 2002, when the last gold mine to shut down in the Western Basin of the Witwatersrand mining area ceased pumping, mine flooding had occurred and AMD began to decant out into the groundwater in the area (Expert Team of the Inter-Ministerial Committee (IMC), 2010, pp. v–vi).

rehabilitation of closed or abandoned mines let alone dealt with the problem of acid mine drainage from those mines.

One of the attendant results of the decline in mining activity and the economic downturn in 2008 was an increase in the outflow, or seepage, of polluted and contaminated water from the old gold mining areas in Gauteng Province, especially the so-called Witwatersrand[9] mining basins[10] of South Gauteng (see map below). Depending on the area, the water may contain toxic heavy metals and radioactive particles (such as uranium). Moreover, this water is dangerous because, if left unpumped, it rises to the surface. This water flow (AMD) out of mines is dangerous for people's health, as well as plants and animals. But the health and environment dangers are not only from the acid mine water, but the environmental pollution includes the dumping of uranium, for many years considered a worthless by-product of gold production, in the huge mine dumps (tailings dumps) and slime dams around Gauteng. For years these dumps and dams have been poorly managed and the rain waters have been allowed to flush all these harmful mining residues downstream and into the underground water table. During the drier winter months in Gauteng, these huge tailings dump stand openly exposed to the winter winds which distribute these carcinogens and heavy metals straight into the air, and they get carried further away, and again rain washed into water sources, thereby entering the water supply system further and further away.

A further AMD exacerbating problem of gold mining in South Africa has been the fact that, while still having 35% of known world gold reserves, most of it is difficult to extract requiring deep-level mining. South Africa, in fact, has six of the ten deepest level gold mines in the world.[11] Many of these deep-level gold mines had to pump out the water that had entered the mines to allow for safe mining conditions. At one stage, more than 120 mines were conducting pumping operations to keep groundwater levels in a mine below the depth of mining operations. But as the South African deep-level gold mines were worked out and abandoned and the subsequent termination of the extraction of underground water from mines, the dewatering of the mines became the responsibility of fewer and fewer mining companies and the mined underground areas (tunnels, drives and shafts) started filling with water, with mines becoming flooded. So, besides the 'usual' accompanying AMD from mining

[9] Literally translated from the Afrikaans as: 'White Waters Ridge'.

[10] The nature of the gold deposits in the Witwatersrand required the development of large, complex underground workings, creating a complex system of underground tunnels and interconnected mines, commonly referred to as 'basins' (The Presidency, Department of Planning, Monitoring and Evaluation, Department of Environmental Affairs and Department Mineral Resources, 2015, p. 4).

[11] These six are: AngloGold-Ashanti's Mponeng Gold Mine (formerly known as Western Deep Levels) is the deepest mine in the world; followed by Driefontein Mine owned by Sibanye-Stillwater, both near to Carletonville; then Kusasalethu Mine on the Far West Rand ten km from Carletonville and the Moab Khotsong Gold Mine—both operated by Harmony Gold, near the towns of Orkney and Klerksdorp; South Deep Mine (7th deepest in the world) owned by Goldfields Mining Company; and Kopanang Gold Mine—also operated by Harmony Gold—being the tenth deepest mine in the world (Anon, 2019).

operations, in the case of the deep-level gold mines in Gauteng, the pumping of water out of these mines to make them safe for mining operations stopped when the mines were closed or abandoned (DWA, 2013).

Consequently, abandoned mines became waterlogged (to the level of the existing water table) by the seepage of water into them and into the tunnels (voids). Thus, in Gauteng Province there was a consequent ever-increasing AMD problem. Since this additional AMD was largely unseen and occurring underground the extent of the AMD only became apparent when it emerged in streams and river systems considerable distances downstream from the actual gold mines (clustered as they were in the Witwatersrand Basin (the urban areas of Johannesburg, East and West Rand regions). Once the AMD reaches the surface and starts to flow downstream into wetlands, streams and rivers, it kills most forms of aquatic life. Since rivers are inter alia used as a water source for the supply of water for agriculture, recreation and drinking purposes, AMD potentially affects the fitness for the use of such water. Much of this AMD seepage flows south into the Vaal River catchment system, which in turn flows into the lower Orange River system—which water system is crucial for agricultural irrigation production for almost 1,000 km downstream (DWA, 2013).

There are also other ancillary consequences to AMD seepage, namely:

- contamination of shallow aquifers;
- if the underground mine water reaches the near-surface environment sinkholes can (and do) occur;
- possibility of AMD impact by resulting water shortages for domestic use, i.e. surface water becomes unfit for human consumption;
- the possible sterilisation of remaining gold reserves in the Witwatersrand Basin (DWA, 2013); and
- AMD flooding may result in inter-mine water migration and thereby threaten neighbouring operational mines, limiting access to economic reefs (Expert Team of the Inter-Ministerial Committee (IMC), 2010, p 2).

3 Looming AMD Disaster Predicted in Gauteng

South African environmentalists have termed Acid Mine Drainage as the "single most significant threat to South Africa's environment" (Naidoo, 2009) and according to Dr Anthony Turton, formerly a scientist at the Council for Scientific and Industrial Research (CSIR) and the director of Touch Stone Resources—a company committed to uplifting South Africans' survival and health prospects—all the gold mining areas in the country, inclusive of the Freestate and Northwest provinces' goldfields, present an AMD threat, but that the "Witwatersrand Basin's massive mine void overlain by dolomite presents the biggest threat" (Turton, as cited in Naidoo, 2009). Turton further likened the South African AMD problem as "…South Africa's own Chernobyl" (Turton, as cited in SABC (Carte Blanche), 2011, np). This AMD was occurring in the most urban, industrial and most densely populated province in South Africa, namely Gauteng with an estimated potentially affected population of five million.[12] In particular many informal communities, dotted all over the Witwatersrand Mining Basin, have made their homes on AMD-contaminated land and feed their animals and crops with contaminated water. Some even making use of mining waste as building materials. By 2014 it was being reported that in some areas of the West Rand there were radioactive lakes. For example, the contaminated water sources in the Krugersdorp Game Reserve had resulted in animals going blind. There were also fears that AMD on the West Rand posed a significant danger to the Cradle of Humankind World Heritage site by hampering efforts to protect the fossil records at and near this site, while on the East Rand, the contaminated water was flowing into a Ramsar[13] registered wetland area (Liefferink, 2014). The government itself recognised that: "…waste from gold mines constitutes the largest single source of waste and pollution in South Africa, and there is wide acceptance that AMD is responsible for the most costly environmental and socio-economic impacts" (The Presidency, Department of Planning, Monitoring and Evaluation, Department of Environmental Affairs and Department Mineral Resources, 2015, p. 4).

But even before 1994 the Apartheid government had been accused of being in collusion with the mining houses and had either denied or downplayed the extent and seriousness of AMD in South Africa (Liefferink as cited in SABC (Carte Blanche), 2011).

Besides the direct effects of AMD on the quality of water resources, there are also indirect but nonetheless as serious, impacts on people's health. The ongoing mining activities combined with the legacy of more than 120 years of gold mining operations have caused water, air and soil contamination in the Witwatersrand region (encompassing as it does the major South African city of Johannesburg, the economic and financial heartland of South Africa). In this region residents have been exposed to "…elevated concentrations of heavy metals and radiation that can contribute to immediate and long-term medical problems ranging from asthma and skin rashes to cancer and organ damage" (Harvard Law School International Human Rights Clinic, 2016, p. 1).

The Department of Water Affairs and Forestry (DWAF) policy document: *The Water for Growth and Development Framework* (WGDF), launched in early 2009, had identified ground- and surface-water pollution, as a result of AMD from abandoned mines, as posing "a threat and an obstacle to securing water for growth and development" (Department of Water Affairs and Forestry (DWAF), 2009, p. 22). The 2009 DWAF WGDF report also stated that in the central and western basins of the Witwatersrand mine systems, this threat was "present and immediate" (DWAF, 2009, p. 22) requiring urgent intervention. But, according to DWAF, the AMD prob-

[12] It is estimated that 25% of the population in Johannesburg (Central Basin) and Ekurhuleni (East Rand region) live in informal settlements, and approximately one quarter of them—400,000 people—are in the mining belt (Tang & Watkins 2011, np).

[13] The Convention on Wetlands, also called the Ramsar Convention, is an intergovernmental treaty that provides the framework for national action and international co-operation for the conservation and wise use of wetlands and their resources (see www.ramsar.org).

lem also had a medium- and long-term scope, with the potential AMD ingress into the water table of the Gauteng Province occurring "for centuries to come" (DWAF, 2009, p. 22).

By the early 2000s, the accelerating closure of uneconomic goldmines led to the decanting (outflow) of AMD from the resulting flooded and water-logged mines. For example, in August and September 2002 on the West Rand, where pumping in most gold mines, with the exception of the so-called Far Western Basin along the Carletonville-Orkney-Klerksdorp axis, had stopped in 1998, the presence of acid mine water outflows near the town of Krugersdorp (Mogale City/Randfontein area of the West Rand region of Gauteng Province) was observed when acidic mine water started flowing from an abandoned mineshaft as a result of the flooding of the mines in this basin to a level where water could flow out onto the surface at a rate of 20-million litres a day. These outflows had begun to pollute surface water in the area, in particular the Randfontein, Tweelopiesspruit and Wonderfontein streams (CSIR, 2009, p. 1; Expert Team IMC, 2010, p. 1; DWAF, 2009, p. 22).

As a result of this outflow on the West Rand the Department of Water Affairs and Forestry (DWAF) issued a directive to the relevant mining companies to treat the affected water to the specified quality before releasing it. But despite some mining companies making significant investments at the affected mines for the upgrading of or installation of new mine water treatment facilities, water quality standards continued not to be met with partially treated water still being discharged into the streams in the area. This had resulted in further contamination of downstream boreholes and other environmental damage, which, if this continued, was likely to result in significant claims for compensation by affected parties (DWAF, 2009, p. 22).

At the time (2002) not much further was done by Mogale City or the government about the AMD in the Western Basin, besides a few cosmetic measures. For example, at the Wonderfontein Stream, the Mogale Municipality had merely erected a warning sign which stated that the water in this stream was "unsafe for human consumption" and that it was not to be drunk or used for cooking purposes (Enviro Editor, 2011b).

The emerging AMD problem was of great concern to several organisations and NGOs, as well as concerned members of the public who had tried for many years to get the government to address this issue. On 10 November 2007, at the inaugural meeting of the Federation for a Sustainable Environment (FSE), it was bluntly stated that: "the government and mining industry have been haphazardly playing with human lives so that they could keep cashing in on our minerals". The feeling at this meeting being that the mining companies and the government had not made any concerted or coordinated effort to fix or deal with the problem of AMD since this "cuts into the profit margins … and that's why they've never done anything about it" with the mines having repeatedly failed in their environmental responsibilities and that "every day that the problem was blatantly being ignored, more damage was being done and more people were getting sick or worse as a result" (Enviro Editor, 2011b).

By 2009, a similar situation as to the AMD mine decanting in the Western Basin near Krugersdorp was developing in the Central Basin (Johannesburg area) and Eastern Basin (Springs-Nigel area) of the Witwatersrand region (East, Central and

West Rand basins—see map above) of the Gauteng Province. By this time acid mine water pollution had reached a crisis point. In the Central Basin, pumping by mining companies had ceased in 2008 (while pumping in the Eastern Basin stopped early in 2011). As a consequence of the absence of operational pumping facilities in the Central Basin, this area, by the beginning of 2009, had flooded to within 900 m of the surface. This was a serious environmental threat to the water aquifer above the underground mines with their rising levels of AMD. While this situation not only represented a potential environmental catastrophe, it was also a threat to the structural integrity of the Johannesburg City Centre (e.g. potential for underground seismic activity, earth tremors and sinkholes occurring which could cause damage to buildings and road infrastructures) (DWAF, 2009, p. 22; Pratt, 2011).

This crisis was simply because some mining companies had abandoned mines without rehabilitating them underground, as well as the above ground mine dumps and slime dams, further resulting in acid mine water flowing into streams, dams and sources of groundwater all over the province. Despite government efforts to get mining companies to take responsibility to 'clean up' mine water seepage, mine dumps and slime dams, in order to contain outflows, as gold became depleted or economically unviable to recover, mining companies abandoned the mines and shut down the water pumps. When the pumping ceased, mines began to fill up with water. Before the mid-2000s, for at least two decades, the government had issued pumping subsidies to mining companies to offset the cost of pumping water that flowed in from adjacent closed mines. However, when all the mines in a goldfield close, no one remains to pump and the government must step in. By 2009 it was estimated that this situation had reached a critical turning point with an urgent need for the implementation of environmentally friendly acid mine drainage control measures before the toxic water under the central Johannesburg city districts reached an environmentally critical and dangerous (to people's general health and clean water standards) level. One of the problems identified in the AMD problem in the Witwatersrand region was the fact that because of the sheer scale of the Witwatersrand gold mining operations there existed a high degree of interconnection between the underground mined areas (i.e., the large voids comprising the access tunnels and mined tunnel areas in these mines) (Expert Team IMC, 2010, p. 19). In fact, in the Witwatersrand Basin there are nearly 800 km of interconnected tunnels and shafts that over the years, have been drilled and dug, the deepest of which is almost 4 km underground. In addition, after more than 120 years of intensive gold mining, the Witwatersrand region has been left with mounds of waste, known as tailings dumps, all underlain by a deep underground network of abandoned mine shafts, which are gradually filling with water. Nearly 400 square kilometres of tailings dumps are scattered in, around and under the city of Johannesburg (Pratt, 2011). All of this is compounded by the fact that many of the AMD problem areas are located in or close to the region's major urban centres, namely: Johannesburg and Soweto complex; Krugersdorp/Randfontein on the West Rand; and Germiston-Benoni-Springs-Nigel on the East Rand—all with a combined population of approximately eight million (population estimate for 2017). As a result, this AMD situation would necessitate large-scale but costly programmes to address the problem of acid mine drainage and prevent further decanting and leakages of AMD into aboveground water resources.

The DWAF estimated the AMD threat to the quality of the water in the water table and river systems of the Witwatersrand region (southern Gauteng) in terms of salinity, levels of sulphates and heavy metals, as being extensive since, in 2008, the AMD discharge from the Witwatersrand basins, while accounting for only 5% by volume of water entering the Witwatersrand catchment area of the Vaal River, produced 20% of the salinity while including toxic heavy metals contained in the AMD mine decanting from the region (DWAF, 2009, p. 22).

But since most of the closed and abandoned mines once belonged to now-defunct mining companies nothing was done about the AMD pollution in the area. Furthermore, the government was slow to respond to what had become a public health issue of great concern to environmental activists. This pollution problem was only addressed, and then in a slow and tardy manner by the government when research reports with the accompanying media publicity began to surface about the extent of the problem all over Gauteng Province. The lack of policy clarity compounded the government's slow response in terms of the delegation of powers between the different levels of government, and between respective government departments at national, provincial and municipal levels. In other words, no one wanted to take responsibility nor 'pickup' the costs of mine rehabilitation. As such, this lack of clarity had led to the fragmentation, overlapping or vague defining of institutional roles and responsibilities. There was also a lack of aligning different legislation (e.g. for mines, water, sanitation, housing and environmental affairs). Furthermore, the problem of discouraging polluters by enforcing the principle of 'polluter pays' was, in the case of AMD pollution, highly problematic given the absence of any responsible organisation (mines closed and abandoned by bankrupt companies) to sanction and fine. Overall, this placed the government in the invidious position of having to fund all rehabilitation costs. In other words, to be reactive rather than proactive (see several environmental reports on AMD from both government departments and NGOs: Auditor-General of South Africa (AGSA), 2009; Expert Team IMC, 2010; Enviro Editor, 2011b; Department of Environmental Affairs and Department Mineral Resources, 2015).

Furthermore, by the end of 2008, the whole rehabilitation programme for abandoned mines of the Department of Minerals and Energy (DME)[14] was in disarray. In February 2008 the Auditor-General of South Africa had conducted a performance audit of the rehabilitation of abandoned mines at the DME. The 2009 Auditor-General's report stated that despite the "…extent of the environmental impact of unrehabilitated abandoned mines, measures were not in place to ensure that abandoned mines were rehabilitated effectively and timeously" (Auditor-General of South Africa (AGSA), 2009, p. 1). Furthermore, as a result, the impact of AMD, both environmental and social, was simply not being addressed. This was compounded by "a lack of accountability, delays with the progress of planned projects, and inefficient service delivery … [as well as] a lack of capacity [in the Department] [which had further] delayed implementation of [mine rehabilitation] projects" (Auditor-General of South Africa (AGSA), 2009, p. 8). The Auditor-General had also found that there was: "…no approved strategy by the Department to guide and

[14] Renamed in 2010 as the Department of Mineral Resources.

inform its [mine] rehabilitation programme" (Parliamentary Monitoring Group (PMG), 2010). Besides the slow pace of existing mine rehabilitation, there had also been an "overt lack of rehabilitation procedures and policies" (PMG, 2010). In addition, in their report the Auditor-General had found that there had occurred unacceptable delays in the implementation of individual mine rehabilitation projects due to tender irregularities with several projects needing to be re-advertised, further delaying the implementation of any rehabilitation. Moreover, the Auditor-General, in its audit of mine rehabilitation projects, found that the DME had under-budgeted or not allocated sufficient funds to the Department's mine rehabilitation programme. As a result, the Auditor-General had been scathing about the absence of project action plans for approved projects accompanied by a crucial lack of departmental oversight over external contractors (PMG, 2010). To ensure that the DME fulfilled its mandate regarding mine rehabilitation responsibilities, the Auditor-General had recommended that DME "capacitate [their] mineral regulation branch" so that the inspectors for regulation compliance could timeously follow up on all mine rehabilitation projects and measure progress and make follow-ups on progress and compliance (AGSA, 2009, p. 9).

In response to the Auditor-General's Report, the Department for Minerals and Energy had stated that: "…historically there were no legislative measures to manage the impact of mining on the environment and the strategy could not be formulated by the Department until the magnitude of the problem was determined by research…[and] high priority sites had been identified" (PMG, 2010). In addition, the Department estimated that its then current liability for mine rehabilitation in South Africa would cost around R30 billion (USD$3.75 billion at 2009 conversion of R8/1$), but its annual budget for mine rehabilitation was a mere R50 million (USD$6.25 million) (PMG, 2010). The Council for Geoscience had also estimated that the building of treatment plants for AMD in Gauteng would cost more than R5 billion (USD$625 million) (AGSA, 2009, p. 5). Such high mine rehabilitation and water treatment costs,[15] even over an extended period, were obviously beyond the financial and economic means of South Africa.

4 Appointment of Inter-Ministerial Committee to Investigate Acid Mine Drainage

As a result of not only the Auditor-General's report but also the news reports of the environmental and ecological crisis emerging on the Witwatersrand region, the government had eventually, in October 2010, appointed an Inter-Ministerial Committee (IMC), comprising of the Ministers of Mineral Resources, Water and Forestry, Environmental Affairs, Science and Technology and the Minister in the Presidency: National Planning Commission, to specifically address the AMD crisis in the Witwatersrand Goldfields Area of the Gauteng Province.

[15] AMD exponentially increases treatment costs since increased loads of discharged effluents increases the associated costs of purifying water for domestic consumption and agricultural use (Makgae, 2012, p. 330).

For several years, environmental monitoring NGOs and activists' blogs had tried not only to alert the public but also galvanise the government and mining houses to address the AMD issue in an organised, sustained and regulated manner. As one environmental activist put it at the time of the IMC reports:

It's a good thing that this issue [AMD] is now out in the public's eye, exposed for all to see and those that have caused this issue (Mining Houses, Government) can no longer run and hide and be ignorant about this very serious issue. But while it's a good thing, it's also not good that this has been kept knowingly under wraps for so many years. These issues have been pointed out to government and the mining industry for years now and those reporting them have been ignored, ridiculed and threatened (Enviro Editor, 2011b, np).

One of the immediate tasks of the IMC had been the appointment of a Team of Experts to undertake research on different aspects of the AMD emerging water crisis on the goldmines. The Panel Team of Experts submitted their first report to Cabinet on 9 February 2011 (DWA, 2017). Among the intervention plans mooted were the following:

1. The development of an Integrated Water Quality Management Strategy;
2. Updating and expanding the monitoring infrastructure for the existing water quality control programmes to more effectively and closely monitor water quality and the rate at which acid mine drainage was rising in each of the Witwatersrand basins (Western, Eastern and Central—see Figs. 12.1 and 12.2);
3. Implement a system to specifically and scientifically monitor the underground seepage (ingress) water levels in closed and abandoned mines and instituting an Early Warning System as part of municipal and regional Disaster Management Plans (in other words monitoring should be continuous on an official organised basis and not left, as in the past, to concerned NGOs and civil society volunteer organisations);
4. Pump water out of the affected gold mines in the Eastern, Central and Western basins to lower the underground mine water levels to prevent further surface seepage (decanting) of AMD. In other words, pumping[16] out water from a point below the old mine workings to prevent it from mixing with shallower, clean groundwater and controlling the flow of water into the mines by capping or sealing openings;
5. Implement a comprehensive treatment programme for pumped mine water (neutralisation by correcting Ph levels, removal of heavy metals) and desalination[17] of AMD by High-Density Sludge (HDS) and Reverse Osmosis (RO) processes

[16] Pumping and managing the influx of water into the mines is also expensive, especially if it needs to be undertaken indefinitely—even if the previous subsidies for pumping directly to mining companies were diverted to the DWA's budget for this purpose. The pump-and-treat programmes are also energy intensive.

[17] The desalinisation process, by means of one of several treatment methods involving chemical precipitation, membranes (such as reverse osmosis), ion exchange or biological sulphate removal, all of which produce clean, potable water, are costly and produce waste products of their own (Expert Team IMC, 2010, p. 74).

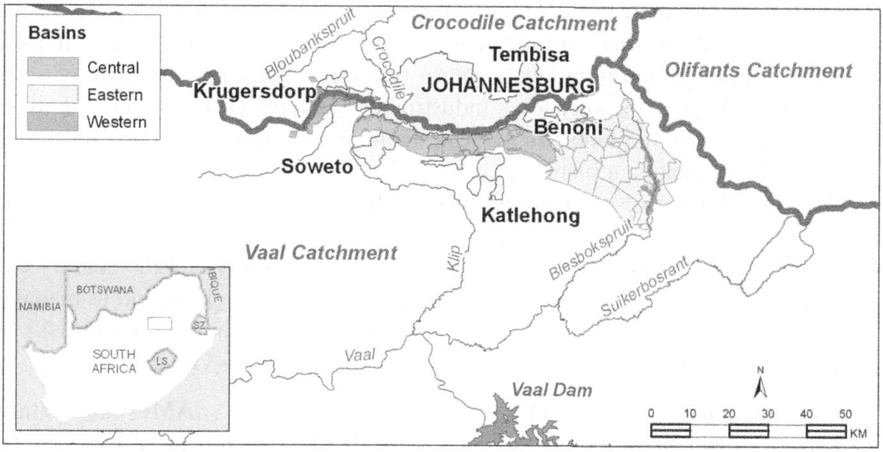

Fig. 12.1 The Witwatersrand catchment area for the Vaal River (The gold mines releasing AMD feed south into the northern Vaal River catchment systems mainly from the Central, Eastern and Western mining basins as indicated in this map). (Source: DWA, 2013)

Fig. 12.2 Merafong City Local Municipality Warning (Source: Enviro Editor, 2011b)

respectively and building of brine evaporation ponds. For further cost savings such treatment plants to be sited at a treatment location situated near the abstraction point before being discharged into surface water resources systems. Treatment plants should also include, because of the high presence of radioactive uranium in the gold-bearing mined ore, the Ion Exchange Process for uranium removal—another costly addition to AMD treatment;

6. To investigate options for supplying to end users of treated water from the neutralisation and desalination treatment sites. In other words, the re-use of treated effluent underground mine water return flows by additionally removal of salts making it fit for, at the very least, industrial use with the eventual aim of domestic consumption, the latter implied the building or expanding additional Rand Water Board treatment plants in Gauteng;[18]
7. Undertake further research to optimise sustainable solutions in the long term; and
8. Devise management solutions to reduce the risk to society and the country for bearing the costs of pumping into perpetuity when mines close, and to manage and control decant on surface of highly polluted water, with its attendant risks to health (DWA, 2013, 2017; Expert Team IMC, 2010, pp. vi–viii & 8).

A final recommendation of the IMC Expert Team being that: "…AMD intervention and management measures are undertaken in the Western, Central and Eastern Basins [of the Witwatersrand region] as a matter of urgency" since the current treatment of the overflowing acid water was totally inadequate (Expert Team IMC, 2010, p. vii). The report had found that the current water qualities in all three priority areas of the Witwatersrand (Eastern, Central and Western basins) were "not fit for untreated discharge to the environment or direct use" (Expert Team IMC, 2010, p. 66). The report further warned that if the water in affected mines was allowed to continue to rise it would soon begin to flow out into low-lying areas all over the Witwatersrand region, even posing a threat to the important tourist attraction on the West Rand, namely: The Cradle of Humankind World Heritage Site (Expert Team IMC, 2010, p. vi & 23). Moreover, the Team of Experts reiterated that in making all their recommendations to government they aimed to "…avert impending crises and stabilise the situation" (Expert Team IMC, 2010, p. viii).

The Minister of Water Affairs, Minister Edna Molewa, was further instructed, as part of the above AMD pumping and treatment action implementation plan, to issue a notice, in terms of Section 19[19] of the National Water Act, to previous and existing mining companies in the Central and Eastern basins, that the mines in these two

[18] But at the time such a scheme to sell the treated water to other companies for industrial use had very little financial incentive for mining companies to treat the mine water as part of a commercial venture, since the cost to treat the AMD water then currently exceeded the price of potable water. In addition, taxpayers, even in a public/private venture of this nature would be reluctant to pick up any costs (Pratt, 2011). In response, the Government set up a public-funded utility, Western Utilities Corporation, with plans to pump the water to a central plant where it could be treated and sold to Rand Water at a profit, thus funding the clean-up process (SABC (Carte Blanche), 2011). But this all still had to be built before such a scheme could become operational.

[19] Section 19 of the National Water Act deals with the prevention and remedying the effects of water pollution by a land user for any activity which "…causes, has caused or is likely to cause pollution of a water resource, must take all reasonable measures to prevent any such pollution from occurring, continuing or recurring". Such person/land user company if they do not comply then the Department "…may recover all costs incurred as a result of" having to implement remedial steps to deal with any resulting pollution due to non-compliance to a departmental directive under Section 19 (Republic of South Africa, 1998. National Water Act 1998, Chapter 3, s19)

basins must not further pollute the water resources as a result of their activities. In respect of the Western Basin mines, where pollution had already occurred,[20] a directive in terms of Section 20[21] of the National Water Act to be issued whereby the DWA would continue to implement remedial interventions and, more importantly would be allowed to claim from the responsible mining companies for the expenses incurred in these remedial processes. Unfortunately, a caveat had been added to this claim directive, namely: "… [recovery of expenses] as far as is possible". This was obviously in recognition that many of the already polluting mines had been closed and abandoned by their mine owners, many of whom had gone out of business due to the prevailing tough economic times or alternately had been forced to cease operations due to owning substantially uneconomic and loss-making mines. So, there was little recourse to getting any one to pay for cleaning up polluted water courses and surface water or getting anyone to institute and pay for expensive preventative measures.

The DWAF Water for Growth and Development Framework Report of 2009 had also previously recommended that directives be issued to mines to develop an Action Plan for mine closure which would include the immediate (upon closure) construction of mid-shaft plugs in these mines as well as the installation of pumping stations in these mines. However, at the time DWAF had recognised that such constructions would be expensive and the monthly maintenance and operating costs would potentially be millions of rands, which was, therefore, not a long-term sustainable solution to the AMD problem. An obvious result of this, especially within the context of mine closure companies themselves closing down, such operating costs then reverting by default to the State and becoming the State's responsibility to continue to fund such pumping and water neutralisation operations. At the time, for the DWAF, a more attractive long-term solution was for the AMD to be converted to potable water and sold to the Rand Water Board (the water utility supplying water to all the municipalities in the Witwatersrand region) with the building of water treatment plants using the most appropriate and cost-effective technologies close to the most heavily affected areas. Such water could then be sold to local municipalities for supplying their residents and to industrial users (DWAF, 2009, p. 40).

But all these plans involved high additional costs. One of the research panels of the IMC Expert Team was mandated to investigate the feasibility of the imposition of an 'Environmental Levy or Tax' (DWA, 2013).

---

[20] One company, based on the West Rand in the Carletonville area, in efforts to reduce the building of costly big tailings dumps and/or slimes dams pumped all the AMD effluent back underground straight into the caverns (voids) they had created in their mining operations. By doing so they simply dumped all the toxic AMD and carcinogen residues straight into the groundwater (Enviro Editor, 2011a).

[21] Section 20 being that section of the Act where pollution has already occurred and ordering the perpetrator(s) to comply with Section 19 and if not then costs for remedial steps undertaken by the Department can again also be claimed from those responsible for the resulting water pollution (Republic of South Africa, 1998. National Water Act 1998, Chapter 3, s20).

Upon receipt of the Experts' Report in February 2011 the government had acknowledged the gravity and "serious challenges" of the AMD situation in the Witwatersrand areas of Gauteng and had accepted that "…it is critical [and] as a matter of urgency" that the pumping and treatment of mine water be implemented forthwith in the Western, Central and Eastern Basins. Accordingly, the Cabinet had approved the setting up, with immediate effect, of a Hydrological Monitoring Committee (HMC) under the chairmanship of the Department of Water Affairs to monitor the re-watering and quality of water emanating from the East, Central and West Rand mining areas. This recommendation included the construction of an emergency treatment plant in the Western Basin to treat uncontrolled AMD partially decants and construction of pumping infrastructure to prevent AMD decants to, at the very least, maintain the agreed upon Environmental Critical Level (ECL) which was set at 150 m below the surface. In addition, to protect the environment in the Central Basin by maintaining water levels at or below the Environmental Critical Level (ECL) of 150 m, pumping infrastructure with the appropriate water treatment plants also to be implemented as soon as possible (which included the refurbishment of the existing neutralisation plant), while the reinstatement of pumping and treatment of mine water in the Eastern Basin to be implemented (Water Affairs Minister, Edna Molewa, also co-chairperson of the Inter-Ministerial Committee, as cited in Government Communication and Information System (GCIS), 2011).

In other words, to pump water out of the mines in order to lower the underground mine water levels, in particular from the Western Basin—the worst affected area having a tributary (Tweelopies Spruit [stream] running into the Klip [Stone] River) of the Vaal River running directly through it. In addition, the pumped underground mine water to be neutralised and the heavy metals removed prior to it being released to surface water resources (DWA, 2017). The Department of Water Affairs made an initial (in April 2011) funding allocation of R225 million (approximately USD$33 million) to implement some of the expert task force's recommendations, including installing the necessary pumps and constructing water treatment plants in each basin, and building infrastructure to transport the treated water (Pratt, 2011).

But while accepting that the partial treatment of acid mine water to neutralise acidity and remove metals was at best a short-term measure, the government had reiterated their stance that in the medium- to long term, it was important that acid mine water be treated to a quality suitable for direct or indirect use (GCIS, 2011, np).

These measures were followed in February 2012 with the initiation of a Feasibility Study (involving not only DWA but also the Council for Geoscience (CGS), the Department of Mineral Resources (DMR) and the Council for Scientific and Industrial Research (CSIR)) to examine long-term solutions to the AMD problem in particular to address the underground mine water induced salt loading of the Vaal River system (DWA, 2013). Again, these costs were for the DWA budget (with smaller contributions from the CGS, DMR and CSIR) and not carried in any way by any of the still operational mining houses. The latter argued that irrespective of their regulated rehabilitation responsibilities towards AMD, they had over the years contributed substantially to the state's finances by means of mining taxes and mineral levies. The mining houses also pointed out that the main culprits of the AMD problem were the numerous closed and abandoned mines and not their still operating gold mines. As an interim measure, the government was able to budget funding to implement (starting in February 2012) some preliminary remediation efforts, which included the pumping of some of the mines most at risk for flooding and performing basic (partial) water neutralisation treatment of the pumped water (Pratt, 2011).

The preliminary results of the Feasibility Report were released on 31 July 2013 and it was reported that the intervention measures (as implemented above) had managed to restrict outflows of AMD, in particular from the Western Basin, inter alia by means of the successful upgrading and operating of three mine water neutralisation plants. Also after an Aquatic Bio-monitoring Assessment had been undertaken in the area (an initiative by the Management Authority of the Cradle of Humankind World Heritage Site, the CSIR and the Mogale City Local Municipality) found that after the AMD neutralisation intervention had been implemented there had occurred an improvement in the water quality in a number of streams in the area, as well as an improvement in the quality of the groundwater. In fact, between October 2011 and February 2013, the water quality in streams above ground had led to the partial return of aquatic life in the Tweelopies Spruit (DWA, 2013, p. 2). These measures proved that, if AMD was properly but continuously treated, its effects could be contained and kept down (below the accepted benchmark ECL 150 m level).

But problems still existed, in particular at the Rand Uranium Mine No. 8 Shaft on the East Rand where only a temporary pump had been installed to increase the volume of water to be pumped from the shaft from the previous 8 million L/day to 24 million L/day. Part of the problem at this mine being that it had been abandoned and then taken over by new owners, intent on restarting the mining operations, but mining only for deep-level uranium deposits. At the time of the report, the new owners were investigating a more cost-effective method of preventing the decanting of AMD from this mine (DWA, 2013, p. 2).

However, during 2013, in the Central Basin (Johannesburg area) a large and expensive reactor plant with pre-neutralisation and sludge-conditioning chambers and a sludge recycling pump station were being constructed[22] to ensure the protection of the agreed Environmental Critical Level (ECL) of water quality required for treated AMD mine water was reached. Also planned for the Central Basin was the drilling of multiple abstraction points (boreholes) throughout the basin so that the sub-basins within this mining basin could be connected and extraction then be done to the new central treatment works. A similar AMD pumping, neutralisation and discharge infrastructure plant was planned for the Eastern Basin for development at the former Pamodzi/Aurora Gold Mine (Grootvlei No. 3 Shaft) (DWA, 2013, p. 3).

The Pamodzi/Aurora Gold Mining Company was an AMD case symptomatic of what was happening in terms of AMD all over Gauteng Province. At the end of 2008, Pamodzi Gold Mining was provisionally liquidated and its shares suspended on the Johannesburg Stock Exchange. But with the suspension of Pamodzi Gold's East Rand operations at the Grootvlei Mine there was a danger that the Grootvlei underground pumping station would be flooded resulting in the release of untreated AMD water on to the surface. If this did occur, it would have far-reaching regional

[22] It only came online in mid-2014.

consequences since the AMD water would pollute the Blesbokspruit River, which was fed water from the adjacent wetland and which eventually entered the Vaal River catchment. Fortunately, as an interim measure, the government, through the DME Mine Rehabilitation Programme, provided a subsidy of R7.5 million (USD$750,000) for the first quarter of 2009 to keep the underground pumping station in operation (Naidoo, 2009). But again, this was an emergency crisis reactive measure using taxpayers' money as a short-term solution, and with the final closure of Grootvlei mining operations by the end of 2010 the government was forced to pick up the costs of pumping and treatment of AMD from the Grootvlei Mine from the beginning of 2011 onwards.[23]

The main issue, was not whether there was an AMD crisis, but rather who would pay for the clean-up and treatment of AMD water? The mooted (by the Team of Experts) of an environmental levy but with the caveat of "on active mines" be considered as a way to fund the clean-up of legacy damage was not a realistic solution to the funding of mine rehabilitation and combatting AMD. After more than a century of gold mining, during which hundreds of companies and their myriad subsidiaries operated in the region, tracking down the parties originally responsible for the acid mine drainage would be nigh on impossible. Even if responsible parties could be identified as a "legacy of lax environmental regulation, especially under apartheid, makes it difficult to hold former mine owners financially responsible for damage caused by mine closure" (Pratt, 2011), while new mine owners now have to carry the cost of 120 years of irresponsible mining; but at the time the gold-mining industry was in decline and these new mining companies did not feel they should pay or could afford the costs of mine rehabilitation, including pumping and water treatment, particularly while much of the AMD was, in fact, coming into their operational mines from abandoned adjacent mines (Pratt, 2011, np).

As Mariette Liefferink, CEO of the Federation for a Sustainable Environment, succinctly summarised the AMD crisis, stating that: "Over 120 years, in more than 120 mines, companies simply mined, they did not address the long-term environmental impacts, especially the impacts on water. They maximised their profits and

[23] The Pamodzi mines was a particularly distressing saga of mine owner exploitation, disregard of legal and regulatory requirements, not least the abrogation of mine rehabilitation responsibilities. In October 2009, after having been placed in liquidation, Pamodzi Gold Mines was taken over by a Black Empowerment Consortium company called Aurora, when a high court-appointed liquidator had given the new mining company Aurora control of its two gold mines—Grootvlei (East Rand) and Orkney (Far West Rand). This despite the new owners having no previous experience in the mining industry, and amidst accusations from the mining industry that the owners were 'given' these mines because of their political connections, namely: managing director being Nelson Mandela's grandson, Zondwa Gadaffi Mandela; the chairman of the board was Khulubuse Zuma, nephew of President Jacob Zuma, while another board member, was Michael Hulley, the personal legal advisor to the president. When they took over at the end of 2009 they promised to resuscitate the mine but by the beginning of 2010 the company had already begun defaulting in salary payments to approximately 5,000 miners and were also accused of asset stripping the two gold mines by selling off equipment and stopping of pumping operations. By the beginning of 2011 all mining operations had ceased at these two mines forcing the government to take over the pumping operations (Pratt, 2011).

externalised the costs" (Liefferink, as cited in Pratt, 2011). Furthermore, Liefferink was of the opinion that: "the taxpayers do not want to pay because the public feels they should not have to pay for pollution that they did not cause and from which they have not reaped benefits or profits" (Liefferink, as cited in Pratt, 2011). Moreover, Liefferink was adamant that the DWA's existing (2012) water treatment programme involving pumping AMD water out of the Witwatersrand Basin, and then partially treating the water by only neutralising it, would not assist in alleviating the long-term challenge of the AMD problem (Solomons, 2015).

However, there was, in some quarters, the view that in the past the government had benefitted from mining via profit-sharing schemes and mining/company taxes, which funded roads, schools, universities and hospitals that over the years have benefitted members of surrounding mining communities and that therefore current mining companies should not have to pay for the damage caused by previous mining companies (Pratt, 2011).

At the beginning of 2015, it was reported that the above short-term solutions had cost the Department of Water and Sanitation (DWS)[24] about R2 billion (approximately USD$175 million), well short of the estimated costs for medium- and long-term solutions to be implemented (Solomons, 2015).

The short-term intervention measures, as outlined above, were only a start on the proposed more comprehensive but therefore more expensive long-term intervention projects. The Feasibility Study had estimated the capital cost of all the envisaged long-term intervention projects in the three mining basins to be approximately R6.66 billion (USD$666 million at March 2012 conversion rate of R10/1$), while an estimated additional annual maintenance and operating cost to be approximately R990 million (USD$99 million) per annum (DWA, 2013, p. 4). All of which placed added strain on the budgets of the Department of Water Affairs and Mineral Resources. But given the global economic crisis at the time, funding of the costs to prevent further Acid Mine Drainage was problematic, and in South Africa, had to compete for limited financial resources with a host of other serious issues then facing South Africa, including high unemployment, infrastructural improvements (provision of clean water, electricity and other services) to the previously neglected townships, HIV/AIDS, rising food prices and persistent energy shortages (blackouts by the main national supplier ESKOM).

Although at the height of the crisis (2011–2012) the government had focused directly on reducing water inflow into the mine voids and treating (partially) the polluted mine water (a delegated responsibility of the Department of Water Affairs), the Department of Mineral Resources was also instructed to develop future management strategies to deal with abandoned, ownerless mines and to close mines in an environmentally sustainable way, especially when they were highly interconnected, as was the case all over the Witwatersrand.

Besides the practical measures of plugging mines, pumping out water from flooded mines to keep water levels below the ECL agreed upon and building more

[24] The DWA's name was changed in July 2014 to the Department of Water and Sanitation (DWS).

water treatment plants, at the height of the crisis in 2011 the government had also been advised by the Federation for a Sustainable Environment (FSE) and other environmental groupings of concerned citizens that they needed to plan and implement other measures to protect the health and wellbeing of citizens exposed to AMD and other toxic heavy metals such as uranium—which stays radioactive for thousands of years—since much of the environmental and conservation damage had already occurred. Among these suggestions being that the government investigate the following non-AMD water intervention measures, but which would still assist to further alleviate the crisis situation, such as:

- to address the root causes of AMD, namely the waste dump sites—waste-rock dumps, the tailings dams, open pits, receptor dams, polluted wetlands and streams, instead of only being reactive and dealing with the flooding of mines by pumping and neutralising outflow waters (Solomons, 2015);
- relocating people living on or near polluted water courses (to be incentivised to do so by the provision of adequate housing in areas with a safer environment);
- those persons living right on top or on the sides of mine tailings dumps, near sludge dams or on the radioactively contaminated ground also immediately be relocated with the contaminated areas being fenced off permanently (i.e. forever);
- this measure to be further reinforced by requiring a larger buffer between new construction and existing mine waste dump sites;
- mining houses to be actively policed monthly or at least every few months by the government so that they are forced to comply to their environmental policy/legislation responsibilities towards communities (Enviro Editor, 2011a; Harvard, 2016, p. 6); and
- due to the widespread contamination from AMD and from tailings dumps/slime dams the government should take steps to remedy the ongoing harm and prevent additional effects in the near term. In other words, they owed the people of South Africa a commitment both financially and politically for the urgent adoption of a coordinated and comprehensive short-, medium- and long-term programme that both "mitigates the effects of mining and helps the country meet its responsibilities under domestic, international, and regional human rights law" (Harvard, 2016, p. 1).

But most of the intervention measures implemented from 2011 onwards were of a short-term nature, and there were also other allied/ancillary problems to the issue of AMD, namely: the mine tailings dumps and sludge/slimes dams, as well as the negative impact on the health of local communities.

5 Desalination, Mine Waste Dumps, Slime Dams and Impact on Community Health

It was only in 2012 that the government took on the primary responsibility for the pumping and treating of the AMD water and were forced then to pick up the costs directly (at taxpayers' expense), having done away with all subsidies for pumping to mine companies themselves. But the water treatment was only a partial and not a complete treatment at that, since the treatment plants only 'neutralised' the water, which process still left "high concentrations of sulphates and other salts that can cause acute health effects and make water unsuitable for such activities as drinking, bathing, washing clothes and watering livestock" (Harvard, 2016, p. 3). Moreover, members of communities surrounding the contaminated areas had "indirectly ingested AMD, especially by eating vegetables irrigated with the polluted water, meat from cattle that have drunk from local waterways and fish from contaminated bodies of water" (Harvard, 2016, p. 2).

As a long-term solution to improving the quality of neutralised AMD water, its desalination, a more thorough but consequently more expensive water treatment process, was needed to improve water quality and to reduce the strain on South Africa's limited drinking water supply, which in turn was also being used to dilute the neutralised water before its release into aboveground water sources. However, it was only in May 2016 that the government announced a plan to construct desalination facilities[25] in Gauteng for that purpose (which were then mooted to be operational only by 2020) (Harvard, 2016, p. 3). While desalination holds out hope for improving overall water its success would be wholly dependent upon the government implementing it timeously, effectively and within budget—something that the environmentalists were sceptical about given the government's past track record in the tardy implementation of planned AMD intervention projects.

Besides implementing desalination, the government also needed to turn its attention to the problem of other major sources of AMD contamination and pollution, namely: runoff after heavy rains and seepage from the mine waste dump sites—the tailings dumps and sludge dams. Similarly to AMD water, these dump sites contain elevated concentrations of heavy metals, including radioactive uranium. In the Johannesburg area alone, there were more than 200 such waste dump sites (Harvard, 2016, p. 4). According to Liefferink of FSE, the Witwatersrand Basin, as a whole, has about 400 km^2 of mine tailings dumps and sludge dams wherein 380 mine waste sites have been identified as potentially having radioactive mine residue deposits. In addition, this basin area has 270, mostly unlined, tailings dams containing an estimated six billion tons of iron pyrite tailings and 600,000 tons of low-grade uranium. The sludge pumped into unlined dams has its heavy metals sink to the bottom and then leaches back underground (Solomons, 2015). All of this potentially a hazardous and health risk to the entire population in the region

Other than the runoff water from these mine waste dump sites there was also the health hazard of windblown contaminated dust blanketing nearby communities, which has led to persistent complaints of the heightened incidence of asthma and breathing difficulties symptoms, particularly among the elderly and children (Harvard, 2016, p. 5).

[25] The Department of Water and Sanitation would have to fund all the costs of building of the proposed Witwatersrand desalination plants to further treat AMD and remove the saline water from the Vaal River catchment system. In 2015 the DWS had estimated it would cost about R10 billion (USD$800 million) to build these plants (Solomons, 2015).

Compounding this situation was the fact that there were numerous informal settlements near or even on top of the tailings dumps or near the sludge dams—the latter often resorted to for water for drinking, washing and irrigating small informal subsistence food gardens. Their location so close to these contaminated sites further endangered the health of residents. So, a further problem of AMD needing attention was the implementation of adequate dust control measures by establishing root-binding vegetation on mine dumps and around tailings dams. Accordingly, the government was also encouraged to totally prohibit the use of mine waste dump sites for such activities as leisure biking (cycles and motorbike) and other recreational activities that lead to the stirring up dust. The absence of such dust-prevention measures was yet another failure by the government in oversight of mining operations let alone enforcing this required rehabilitation measure on closed mines on mining companies (Harvard, 2016, pp. 5–6).

6 Concluding Remarks

The gold mining industry in South Africa has, for more than a century, contributed substantially to the prosperity of the country, and was one of the main triggers, not only of significant economic growth, but also of the start of industrialisation. In contrast, with the abandonment of gold mines in the Witwatersrand Basin, the resulting AMD has led to considerable environmental degradation and water pollution, not to mention its negative impact on the health of residents in the AMD affected areas.

Unlike the US, especially from the early 1980s onwards, who were prodded by early and growing environmental consciousness backed up by the progressive adoption of anti-AMD methods in terms of prevention, prediction and remediation with a comprehensive and strict regulations approach, South Africa has struggled to enforce even the existing regulations for mine rehabilitation in terms of AMD. This was compounded by the inability to hold any gold mining company in Gauteng liable for costs or even imposing any sanctions/fines let alone compelling them to implement any or all anti-AMD measures ranging from pumping, extracting heavy metals to purifying polluted water, let alone undertaking the rehabilitation of already degraded environments. This was simply due to the fact that gold mine owning companies of abandoned mines were no longer in existence (gone out of business/bankrupt), and therefore could not be sued or compelled to institute these measures or pick up the costs involved, not only of continuing with pumping but also doing rehabilitation work on the closed mines. In fact, there was not one case launched against any mining company with the exception of the Department of Mineral Affairs instituting orders to that affect but only on still-operating gold mines. That action being only to enforce the mine rehabilitation regulations on such operating mining companies (i.e. they might still be operating other gold mines but closing only the most uneconomical ones).

As a consequence, the government (and taxpayers), faced with the inevitable AMD crisis (2009–2011), were forced, in the absence of any operational mines, particularly in the Central Basin, to pick up all costs (developmental and continuation of pumping and treatment) to, at the very least, prevent an environmental and conservation disaster of extensive magnitude. Many of the gold mines were abandoned prior to the government becoming aware of the emerging crisis (in 2008–2009) or were considered insolvent before the full environmental impact of their mining operations became evident. As a result of this, there were many untraceable and literally 'ownerless' mines in the Witwatersrand Basin or new owners refused to take responsibility for the previous owners' actions. Consequently, the government, by default, became responsible for addressing the immediate challenges and associated impacts of AMD, while also having to plan for and institute more permanent long-term measures to address AMD and its associated contamination and pollution effects.

Since 2011, while the South African government has taken several urgent and situation enforced intervention measures to address the adverse AMD impact of gold mining, it has consistently failed to live up to many relevant human rights and other constitutional obligations.[26] At the height of the AMD crisis the government's response was generally slow and inadequate (under the circumstances of the time being more reactive crisis management than proactive). As a result, gold mining in the Witwatersrand Basin created not only environmental, conservation and health risks, but it also prevented community members from attaining numerous constitutionally-enshrined human rights, such as access to clean water and air; safe housing; sustainable and safe agricultural pursuits; non-exposure to hazardous/radioactive materials, contamination or pollution—in other words being protected from all these AMD risks and being provided with a healthy environment in which to live. In addition, the state bears primary responsibility under human rights law for preventing the infringement of any of its citizens rights and what was crucially needed as the AMD crisis erupted was for the government to make a stronger connection between dealing comprehensively in the short-, medium- and long-term with all the facets of AMD and protecting the environment and human health.

The Harvard Law School International Human Rights Clinic (IHRC) conducted on the ground investigations in South Africa in 2010, 2012 and 2014, and followed these up in 2015 and 2016. The IHRC found that the South African government had failed to live up to many relevant human rights obligations and that its response to the AMD crisis in the Witwatersrand gold fields (Western, Central and Eastern basins) had been generally slow and insufficient. As a result, gold mining in Gauteng

[26] Section 24(a) of The Republic of South African Constitution (1996) enshrines the right of everyone "to an environment that is not harmful to their health or wellbeing"; while Section 24(b) states that everyone has the right: "to have the environment protected, for the benefit of present and future generations, through reasonable legislative and other measures that (1) prevent pollution and ecological degradation; (2) promote conservation; and (3) secure ecologically sustainable development and use of natural resources while promoting justifiable economic and social development." (The Republic of South Africa, 1996).

has not only created environmental and health risks, but it has also prevented community members from realising numerous human rights (Harvard Law School International Human Rights Clinic, 2016).

In 2014 there was a concerted effort ("…decrying [the] government for applying solutions to the problem that are inadequate" (Liefferink, 2014), by a consortium of environmental and environmental justice NGOs,[27] both local and international, to hold the State liable for enforcing all anti-AMD measures. Their efforts were based around intensifying a series of 'alternative' non-state strategies, inter alia the following: (1) continual lobbying of government and state departments responsible for dealing with the AMD problem; (2) funding of further independent research for the creation of alternative (to government-funded studies) reports and environmental impact studies; (3) sharing of information/knowledge with other similarly AMD-affected areas in other countries and learning from 'best practices/new scientific findings' in dealing with AMD that might have been implemented in those jurisdictions; (4) engaging with the media (i.e. media-based activism) to better inform the public on the AMD problem and linked issues, for example: water access rights and entitlements; (5) once the impact studies have been done to mobilise affected communities—often those in informal settlements and too poor to exercise any legal options—to register in courts for reparations/compensation once impact studies had quantified damages due to affected communities (in terms of impacts of/on health, contaminated/polluted water, agriculture/farming, biodiversity loss (e.g. wildlife), food insecurity (crop damage); etc.); (6) such mobilisation to be evidence-based by means of the research knowledge development and thereby to encourage a network of activists/NGOs to launch collective action to compel the government to implement long-term sustainable options to solve the AMD crisis in Gauteng Province but also to implement measures to effectively deal with the existing negative impacts on communities and the environment of the AMD contamination and for the prevention of any further pollution from AMD and leakages from existing tailings dams; (7) all of the above to be reinforced by a public campaign ranging from complaint letters directly to officials; petitions and other public campaigns such as street protests/marches—all to strengthen the participation of civil society in dealing with the AMD crisis (Liefferink, 2014).

The consortium of Environmental Justice Organisations (EJOs) also demanded that the government begin to 'get serious' about implementing environmental improvements by hastening the rehabilitation/restoration of AMD-affected areas. Furthermore, that institutional changes be made by drafting new legislation. To investigate additional technical solutions to improve the existing measures against AMD based on a new Environmental Impact Assessment Study (Liefferink, 2014).

Obviously, there is a need to safeguard the purity and quantity of water against irresponsible mining operations, wherever they occur. Such irresponsible mineral development can result in a reduction of the quality of water, through increased pol-

[27] Among these being the Federation for a Sustainable Environment (FSE); Earthlife Africa—Johannesburg; and the Mine Water Action Group (Liefferink, 2014).

lution, sedimentation loads and toxicity levels, leading to a reduced quantity of water being available for use by current and future generations. This falls in line with the principal of sustainable development. There is, therefore, a need to ensure that the best anti-AMD and water pollution prevention strategies should be employed, especially in cases where the environmental risks can be managed.

The problem of Acid Mine Drainage, the attendant water pollution and allied contamination effects was, therefore, not only a mining problem but also an assault on clean water and health rights of citizens living in the Gauteng Province. As such it represented not only a 'water crime' but also an environmental (water and air pollution) and conservation crime (non-rehabilitation and control of the problem) in terms of protecting, conserving and advancing efforts to ensure 'clean' water and air and a healthy living environment to all communities in the Gauteng region.

References

Anon. (2019). The top ten deepest mines in the world. *Mining Technology*. Retrieved October 10, 2019, from https://www.mining-technology.com/features/feature-top-ten-deepest-mines-world-south-africa/

Australian Government. (2016). *Preventing acid and metalliferous drainage: Leading Practice Sustainable Development Program for the mining industry*. Retrieved October 8, 2019, from https://www.industry.gov.au/sites/default/files/2019-04/lpsdp-preventing-acid-and-metalliferous-drainage-handbook-english.pdf

Council for Scientific and Industrial Research (CSIR). (2009). *Acid Mine Drainage in South Africa*. Briefing Note 2009/02 (August). Retrieved August 10, 2016, from https://www.environment.co.za/documents/acid-mine-drainage-amd/AMD-Acid-Mine-Drainage-South-Africa-CSIR-draft.pdf

Department of Water Affairs (DWA). (2013). Feasibility Study for a Long-Term Solution to address the Acid Mine Drainage associated with the East, Central and West Rand Underground Mining Basins, Gauteng Province. *DWA Newsletter, 3*. Retrieved August 2, 2017, from http://www.dwa.gov.za/Projects/AMDFSLTS/Documents/AMD-FS_Newsletter_Edition-3_High-Res.pdf

Department of Water Affairs (DWA). (2017). *Witwatersrand, Gauteng: Acid mine drainage: Long term solution*. Retrieved June 20, 2017, from http://www.dwa.gov.za/Projects/AMDFSLTS/default.aspx

Department of Water Affairs and Forestry. (2009). *The Water for Growth and Development Framework* (Version 7). Retrieved July 7, 2017, from http://www.dwa.gov.za/WFGD/documents/WFGD_Frameworkv7.pdf

Department of Agriculture, Water and the Environment, Australian Government. 1999. *Environment Protection and Biodiversity Conservation Act 1999*. Canberra, ACT: Australian Government. Retrieved October 8, 2019, from https://www.environment.gov.au/epbc

Earthworks. (2017). Acid Mine Drainage. *Earthworks*. Retrieved July 5, 2017, from https://www.earthworksaction.org/issues/acid_mine_drainage/

Enviro Editor. (2011a). *Acid Mine Drainage (AMD) South Africa* (25 February). Retrieved June 7, 2017, from https://www.environment.co.za/poisoning-carcinogens-heavy-metals-mining/acid-mine-drainage-amd-south-africa.html

Enviro Editor. (2011b). *Acid Mine Drainage Threat – Johannesburg, West Rand, Cradle of Humankind, 1 March*. Retrieved June 7, 2017, from https://www.environment.co.za/acid-mine-drainage-amd/acid-mine-drainage-threat-amd-johannesburg-jhb-west-rand-cradle-humankind.html

Expert Team of the Inter-Ministerial Committee (IMC). (2010). *Mine water management in the Witwatersrand gold fields with special emphasis on acid mine drainage*. (Report to the inter-Ministerial Committee on Acid Mine Drainage. Expert Team of the Inter-Ministerial Committee under the Coordination of the Council for Geoscience. December). Retrieved July 27, 2017, from http://www.dwa.gov.za/Documents/ACIDReport.pdf

Foundation for Water Research. (2004). *Neutralization of acid mine water and sludge disposal* (Report No 1057/1/04). Marlow: Foundation for Water Research. http://www.fwr.org/wrcsa/1057104.htm Retrieved June 24, 2017.

Government Communication and Information System (GCIS). (2011). *Acid Mine Drainage Report released by Department Water Affairs – South Africa*. (Media Release, 24 February). Retrieved September 10, 2017, from https://www.environment.co.za/poisoning-carcinogens-heavy-metals-mining/acid-mine-drainage-report-released-by-department-water-affairs-south-africa.html

Harvard Law School International Human Rights Clinic (IHRC) (2016). *The cost of gold: Environmental, health, and human rights consequences of gold mining in South Africa's West and Central Rand*. (October). Retrieved October 10, 2017, from https://www.environment.co.za/wp-content/uploads/2016/10/Cost-Of-Gold-Mining-South-Africa-Acid-Mine-Drainage-AMD-Environmental-Disaster-Harvard-Report.pdf

Jennings, J. R., Neuman, D. R. & Blicker, P. S. (2008). *Acid Mine Drainage and effects on fish health and ecology: A review* (Report prepared for the U.S. Fish and Wildlife Service. June). Bozeman, MT: Reclamation Research Group. Retrieved July 2, 2017, from https://www.reclamationresearch.net/publications/Final_Lit_Review_AMD_08-22-08.pdf

Liefferink, M. (2014). Acid mine drainage, South Africa. *Environmental Justice Atlas*. Retrieved October 8, 2019, from https://ejatlas.org/conflict/acid-mine-drainage-south-africa

Makgae, M. (2012). The status and implications of the AMD legacy facing South Africa. *Proceedings of the International Mine Water Association Symposium*, 29 September–4 October (pp. 327–334): Bunbury.

Minerals Council South Africa. (2018). *Facts and figures 2018*. Johannesburg: Minerals Council South Africa. Retrieved October 8, 2019, from http://www.mcsa-facts-and-figures-september-2018.pdf

Naidoo, B. (2009). Acid mine drainage single most significant threat to SA's environment. *Mining Weekly*, (8 May). Retrieved July 10, 2017, from http://www.miningweekly.com/article/acid-mine-drainage-single-most-significant-threat-to-sas-environment-2009-05-08/rep_id:3650

Ochieng, G. M., Seanego, E. S., & Nkwonta, O. I. (2010). Impacts of mining on water resources in South Africa: A review. *Scientific Research and Essays, 5*(22), 3351–3357.

Parliamentary Monitoring Group (PMG). (2010). *Auditor-General Performance Audit Report on Rehabilitation of Abandoned Mines*. Meeting of the Parliamentary Portfolio Committee on Public Accounts, 31 August. Retrieved August 31, 2017, from https://pmg.org.za/committee-meeting/11951

Pratt, S. E. (2011). All that glitters... Acid mine drainage: The toxic legacy of gold mining in South Africa. *Earth*, (September). Retrieved August 24, 2017, from https://www.earthmagazine.org/article/all-glitters-acid-mine-drainage-toxic-legacy-gold-mining-south-africa

Republic of South Africa. (1998). *National Water Act* (Act No 36 of 1998). Retrieved June 20, 2017, from http://www.dwaf.gov.za/Documents/Legislature/nw_act/NWA.htm

SABC (Carte Blanche). (2011). *Carte Blanche story on acid mine drainage* (25 February). Retrieved July 14, 2017, from https://www.environment.co.za/poisoning-carcinogens-heavy-metals-mining/carte-blanche-story-acid-mine-drainage-amd.html

Solomons, I. (2015). Questions raised about govt's approach to acid mine drainage. *Mining Weekly*, (23 October). Retrieved August 27, 2017, from http://www.miningweekly.com/article/change-urged-2015-10-23

South African Government. (2019). *Minerals*. Retrieved October 5, 2019, from https://www.gov.za/about-sa/minerals

The Auditor-General, South Africa. (2009). *Report of the Auditor-General to Parliament on a performance audit of the rehabilitation of abandoned mines at the Department of Minerals and Energy* (October). Retrieved July 24, 2017, from https://www.environment.co.za/Auditor-General-report-on-rehabilitaton-of-abandoned-mines-at-the-DME.pdf

The Presidency, Department of Planning, Monitoring and Evaluation, Department of Environmental Affairs and Department Mineral Resources. (2015). *Report on the implementation evaluation of the effectiveness of environmental governance in the mining sector* (11 August). Retrieved September 15, 2017, from http://www.Environmental-Gov-in-the-Mining-Sector-201525-FINAL-Report-2015-2008-2011-JB.pdf

The Republic of South Africa. (1996). *Constitution of the Republic of South Africa 1996*. Retrieved October 5, 2019, from http://www.justice.gov.za/legislation/constitution/SAConstitution-web-eng.pdf

Tang, D. & Watkins, A. (2011). Ecologies of gold: The past and future mining landscapes of Johannesburg. *Places Journal*, February. Retrieved October 10, 2019, from Available at: https://placesjournal.org/article/ecologies-of-gold-mining-landscapes-of-johannesburg/

United States Environmental Protection Agency (EPA). (2019). *Summary of the clean water act*. Retrieved October 5, 2019, from https://www.epa.gov/laws-regulations/summary-clean-water-act

United States Geological Survey. (2000). *Mine drainage*. Retrieved October 5, 2019, from http://geology.er.us.gs.gov/eastern/environment/drainage.html

United States Geological Survey. (2019). *World gold production by country* (2004–2018). Retrieved October 8, 2019, from https://www.usagold.com/cpmforum/world-gold-production-by-country/

Chapter 13
Renegotiations of Privatization Agreement of Water Utilities: Anecdotal Evidence from European Union

Katarina Zajc

1 Introduction

The aim of this chapter is to examine, in a couple of case studies, the cause of the renegotiation and/or termination of the water infrastructure concession agreements and any other privatization contract, including outright asset sale, in developed countries, focusing on certain privatization agreements in the water sector in EU. Due to the nature of the water sector, with, upfront, high and sunk investment, negative and positive externalities on health and growth of the regions or countries, the choice between the ownership of water assets, private or government owned, is not obvious and neither is the type of privatization contract, which would bring efficient supply of water if privatization is the preferred option (Zajc, 1997, 2004). However, we should expect that concession agreements should be similarly efficient and/or renegotiated in EU member states due to the similarity of institutional environment.

Water asset privatization and municipalization or nationalization is a highly debated topic in the last 30 years.[1] Even though historically water distribution and treatment was in private hands, certain countries and regions provided water by public entities for a very long time (Foreman-Peck & Millward, 1994).[2] However, increased scarcity, alleged inefficiencies in the provision of water and municipal budget constraints motivated the privatization processes in the water sector. After a

[1] Even though this chapter talks about water, it is not concerned with a water as an economic good and pertaining rights, but with the water utilities and assets, and distribution of water to end users.

[2] In the UK, for example, the water sector was unregulated until about 1860. After 1860 the government started to regulate the water sector, and around the time of Second World War the infrastructure was nationalized. Water supply utilities were also nationalized, even though 28 small water utilities were also nationalized, even though 28 small water utilities stayed private.

K. Zajc (✉)
Law Faculty, University of Ljubljana, Ljubljana, Slovenia
e-mail: katarina.zajc@pf.uni-lj.si

K. Eman et al. (eds.), *Water, Governance, and Crime Issues*,
https://doi.org/10.1007/978-3-030-44798-4_13

wave of privatization in the 1990s, the question arose under which conditions privatization brings about increased efficiency of water distribution and better water quality, together with expanded service and payment discipline. Proponents of privatization, based on empirical evidence, claim that privatization of water services was a success, opponents, on the other hand, also based on empirical evidence, suggest that privatization was not successful and that privatized assets should be and are nationalized and are returning back to the municipal hands (Andrade, Cruz, & Saramento, 2018). Kishimoto, Lobina, and Petitjean (2015) report that 235 cases of remunicipalization occurred in 37 countries from March 2000 to March 2015, regardless of institutional development and despite considerable effort of donors to promote privatization. The push for privatization comes mainly from discontent with privatized water supply and dissatisfied local governments. The reasons for discontent are many, for example poor operational performance, under-investment, disputes over operational costs and price/tariff increases, soaring water bills, difficulties in monitoring private operators, lack of financial transparency, workforce cuts, and poor service and water quality (Lobina, 2017). Currently, there exists only around 10% of the private water supply in the world (Getzner, Köhler, Krisch, & Plank, 2018).

One of the weaknesses of privatization is also high degree of renegotiations of privatization agreements, namely concession agreements (or in some cases, lease agreements) and regulation of outright sales.[3] The problem arises since through renegotiations either the private sector, or the government can change the terms of concession agreements in their favour, which is not necessarily increasing the social welfare and therefore undermine the initial goal of privatization in the first place.

The hypothesis stated in Zajc (1997) still holds that in the water sector the more developed the institutions, the less privatization agreements should be renegotiated or terminated, since the government intervention and private sector intervention and renegotiation of privatization contracts is costlier for both the more the institutions are developed since they both might loose much more credibility and renegotiation is therefore less desirable. It follows then that the concession agreements, or better yet, all privatization agreements, would be subject to more renegotiations in countries with weaker institutional environment, since the motivation for the breach or to get gains that are outside the agreement are costing parties less in the weak institutional environment. Therefore, there should be no difference in renegotiations or termination of private sector supply and distribution of water in countries, which are roughly at same level of institutional environment. However, we see in the past 20 years, even in developed countries, that concession agreements are renegotiated everywhere, even in countries with high level of institutional development. After examining a few of case studies of water privatization contracts in EU countries, the chapter discusses whether there are any parallels between the renegotiation of concession agreements in developed or developing countries. The presented case

[3] Even though lease and concession agreements are legally not comparable in its entirety, for the sake of this chapter they will be treated as the same legal institute, since form the privatization point of view they deal with the same legal problems to a considerable degree.

studies show us that even in developed institutional environment renegotiations are reoccurring, it seems that mainly due to some local weaknesses of the institutional environment and economic variables, such as variable demand for water. However, the anecdotal evidence also shows that the provisions in concession agreements matter. The conclusion is that improvement of local institutional environment before starting with the process to confer the rights to water concession and following the good practices with respect to provisions of the concession agreements is a prerequisite for improvement privatization contracts in water sector.

The chapter is structured as follows. After the introduction, the second part overviews the model of privatization of water utilities and institutional environment with special emphasis on the renegotiation of privatization contracts after a brief overview of two cases of renegotiations of water concessions outside Europe which are representative of the failures and renegotiation of the water privatization contracts. The third section presents four cases of water asset and distribution privatization contracts in EU, namely Bucharest, Grenoble, Paris and concession agreements in Portugal. Discussion and conclusion follow.

2 Overview of the Literature on Renegotiation of Privatization Contracts and Institutional Environment

Guasch (2004) reported that in water sector in Latin America between the years 1985–2000, renegotiations affected 74% of concession agreements and renegotiations have occurred on the average 1.6 years after the award, despite the fact that most contracts were signed for more than 15 years. Probably the most famous renegotiations and finally cancelation of water concession agreements occurred in Buenos Aires, Argentina and Cochabamba, Bolivia. In May 1993, a 30-year water concession was granted to a private company in Buenos Aires after an aggressive bidding process among three consortia that prequalified for bidding. The concession was awarded to the bidder that offered the lowest tariffs.[4] The winner (a consortium) got the water infrastructure in very bad shape, tariffs were extremely low (even though the local government increased them before privatization took place), many people did not have access to drinking water, despite abundant sources of water, and water leakages were extremely high (Zerah & Graham-Harrison, 2001). The ground for privatization was prepared before the concession was granted with the hike in tariffs beforehand, and the government wrote off old debts of the public company that supplied water before and reduced its workforce. The goals for the concessionaire were extremely ambitious, expanding water network, reducing water leakages, which amounted to the required investment in the vicinity of US$4 billion (Schiffler, 2015).

[4] According to sources (Zerah & Graham-Harrison, 2001), outright asset sale could have posed legal problems and therefore the government did not want to pursue the outright sale.

After the award of the water concession, mainly those consumers that were already connected to the water system, benefited from quality and reliability of service. However, poor households, which were mainly without connections before the award of the concession, did not benefit at all since the water system either did not reach them in the years after the concession was awarded or if it reached them the costs of connection, which was mandatory, was too high (US$44 per month for 2 years and the average monthly income in poor neighbourhoods was US$220/month) that they were not able to afford it an avoided the connections at all costs. Even though a regulative agency was established in 1993 (ETOSS—Ente Tripartite de Obras de Servicios de Saneamiento), the staff was poorly qualified to tariff setting and had no experience in regulation. In addition, the concession agreement allowed ETOSS to intervene in operational decisions of concessionaire. The first renegotiation occurred only a year into the concession agreement in 1994 when a 13.4% tariff increase was approved in exchange for increased investments. Due to high infrastructure charges to connect new users, many of them were either not paying their bills or refused to connect and by the end of 1996, Aguas Argentina, a winning consortium headed by the Lyonnaise des Eaux-Dumez,[5] was owed US$30 million, and their revenues were lower than projected at the signing of the concession agreement by US$217 million mainly due to non-metered water, refusals to connect to water network and/or pay for it and non-payment of invoices for supplied water, which made a considerable dent in their budget. Renegotiations started in 1997, but were bypassing ETOSS, and the concessionaire negotiated directly with the federal government, which granted them higher tariffs, decreased investment targets, abolished penalties for not meeting the investment targets and also implemented a lump sum environmental and connection charge for connected users, which the concessionaire received even before new connections were built. The overall increase in tariffs was hard to determine exactly since some of the revenues and benefits received in renegotiations did not arise in higher tariffs. Tariffs were again increased by 4.6% in 1999. Afterwards with the financial crisis in Argentina, which started in 1998, everything spiralled downward. In March 2006, the concession agreement was rescinded. The water supply reverted back to the public company, which was struggling since the tariffs were frozen for 10 years until a 300% increase in tariffs were allowed in December 2011 (Schiffler, 2015). Aguas Argentinas sued the Argentinian government after the rescission of the concession agreement and the International Center for the Settlement of Investment Disputes of the World Bank group decided that the government breached the concession contract but did not determine the amount of damages and deferred the award of damages to an independent expert (Schiffler, 2015).

The concession agreement in Cochabamba, Bolivia, was even more-short lived. As a city positioned at more than 2,500 m above sea level, fresh water is not abundant and the public sector did not do a very good job at delivering water since only

[5] Now part of Suez, as a result of 1997 merger between the Compagnie de Suez and Lyonnaise des Eaux-Dumez.

60% of the population was connected to the network, water was available only four hours a day, water quality was poor and its performance lagged behind the performance of other Bolivian utilities (Schiffler, 2015, p. 18). At the deadline of the tender for very ambitious concession there were no bidders. However, Aguas del Tunari submitted its bid after the deadline and was consequently awarded the 40-year concession contract in September 1999. The award of concession coincided with the passing of the law which gave an exclusive water supply rights to water concession holders, therefore Aguas del Tunari and the decision of the local government to increase water tariffs, which were requested by public authority supplying water before the concession agreement was in effect (Nickson & Vargas, 2002). Increased tariffs and monopoly on water supply angered citizens of Cochabamba and the protests against privatization started in January 2000. In April of the same year, the concession agreement was rescinded and the water supply was in the hands of local government once again (Cochabamba water privatization, 2019). Aguas del Tunari claimed damages from Bolivia, but later on dropped the claims in the amount of US$25 million. The public water supply did not improve since reclaiming its function in year 2000 (Schiffler, 2015).

These two biggest failures in granting concession agreements or perform outright privatization show us that institutional environment and nature of concession agreement play a big part in granting and renegotiating the privatization agreements. As stated in Zajc (1997, 2004), the one sided, two-stage prisoner's dilemma for water assets privatization predicts that the private sector and the government will cooperate in reaching a certain type of privatization contract, concession or outright sale of assets, depending on the type of privatization contract and the level of institutional environment. The type of privatization contract is important since in the concession agreement the property rights are divided. The government keeps the control rights (also in terms of regulation) and the concessionaire obtains the right to the residual stream of profit. In asset sale, rights are not divided to the same extent, even though outright privatization is still subject to regulation, which might divide the control rights and the right to residual stream of profit. Also, both types of contract differ in their continuity of the relationship with the government. This continuity influences the reputation of contracting parties, renegotiation over time and the transparency of monitoring and regulation over time. Property rights theory (Alchian, 1965; Alchian & Demsetz, 1972; Grossman & Hart, 1986; Hart & Moore, 1990; Klein, Crawford, & Alchian, 1978) combined with the theory of the firm (Coase, 1937) and transactions costs theory (Williamson, 1987) imply that when division of property rights is costly, one party should hold all the property rights. However, the prediction of the two stage one-sided prisoner's dilemma privatization model (Zajc, 1997, 2004) is that the concession agreements, and therefore the efficient and good quality water supply, will be more successful in strong institutional environment and therefore asset sale is more desirable in less than ideal institutional settings. In other words, concession agreements are more at risk of renegotiations in countries with weak institutional environment. The model predicts that concession agreements and assets sale, on the other hand, should be equally successful in strong

Table 13.1 Significant determinants of renegotiation (Guasch & Straub, 2006)

	Government-led renegotiation	Firm-led renegotiation
Existence of a regulator	Negative	Negative
Price cap regulation	Positive	Positive
Duration	Positive	Positive
Investment requirements	Positive	Negative
Exclusive private financing	Negative	Positive
Quality of bureaucracy	Negative	Negative
Corruption	Positive	Negative
Election (lagged)	Positive	Positive
Growth (lagged)	Negative	Negative
Minimum income guarantee	Positive	Positive

institutional environment (Zajc, 2004).[6] It follows that concession agreements (or asset privatization) should be subject to more renegotiations and termination in countries with weak institutions and not so much in countries with strong institutional environment. However, the anecdotal evidence presented in this chapter does not support this claim (Table 13.1).

Laffont (2003), Guasch, Straub, and Laffont (2003), and Guasch and Straub (2006) introduced a model of renegotiation of concession agreements in which incompleteness of the contract and therefore motivation for the renegotiation is linked to contract characteristics, regulatory environment and economic shocks. The renegotiations are led either by the private sector or the government. They used a dataset of 307 concession agreements in Latin America between 1989 and 2000 in water and transport sector in order to empirically test the impact of different variables on the probability of either the government or private sector led renegotiations of the concession agreements, since 167 out of 307 concession agreements were renegotiated during the life of the contract. In 53 incidences the renegotiations were initiated by the private sector, and in 94 incidences they were initiated by the government sector, in 15 incidences both, the government and the private sector agreed on starting renegotiations. The variables that authors found to be significant for renegotiation were the existence of a regulator at the time the concession is signed, type of financial regulation of concessionaire, namely price cap regulation, duration of concession agreements, investment requirements during concession agreements, exclusive private financing, quality of the bureaucracy, existence of corruption, effect of elections, growth of the economy and minimum income guarantees. The empirical results showed the impact of the mentioned variables on the probability of the renegotiation of the concession agreements, initiated either by the government or the private sector.

[6]The strength of the institutional environment is measured by the credibility of government and independence of judiciary, which is in turn measured with different indices, for example Property Rights Index, Index of Economic Freedom, Regulation and Corruption Index, Economic Risk Index, Doing Business.

The empirical results in both studies Guasch and Straub (2006) and Guasch, Laffont, and Straub (2007) show that the existence of regulator has a negative impact on the probability of renegotiations for either government or private sector, so does the economic growth and the quality of bureaucracy, which means that the higher the quality of the bureaucracy or the better economic conditions, the better the institutional environment and less motivation to start renegotiations either by the government and private sector. Price cap regulation, duration of the contract, elections and minimum income guarantee for the concessionaire have positive impact on the probability of renegotiations by both, the government sector and private sector. However, three variables have a different motivation for the government and the private sector to initiate the renegotiations, namely the level of corruption, whether the investments are financed by public or private funds and whether there are investment requirements in the concession agreement. Corruption and investment requirements have a positive impact on the probability of concession renegotiations for government and negative impact on the probability of starting renegotiations by the private sector. For example, in more corrupt environments the private sector has less motivation to start renegotiating the concession agreement and the government is more likely to initiate the renegotiations. Exclusive private financing on the other hand, has a positive impact on probability of renegotiation for private sector and negative impact of the on the probability of renegotiation by the government.

Empirical results therefore support the claim that the quality of institutional environment, namely existence of regulatory agency, existence of corruption, quality of bureaucracy, and political cycle (elections) are significant variables that impact the probability of renegotiation of concession agreement. The same goes for variables that arise out of the division of right, namely price cap regulation, level of investment and income guarantee. As pointed out by empirical results, economic shocks also impact the probability renegotiations, since they influence the economic variables agreed on in the concession agreement and/or the variables determined by regulation.

3 Europe: Privatization and Remunicipialization of Water Assets

This section of the chapter reviews four privatization agreements and country overview and their renegotiations and/or rescissions. Privatization agreements were concluded in Europe, and most of the analysed countries were a part of EU if not at the beginning of the concession contract, but during the concession agreement. The section starts with the analysis of the Bucharest water concession, then it reviews agreements in Grenoble, Paris and concludes with privatization of water distribution system in Portugal. The presented anecdotal evidence mostly confirms the above predictions and empirical results with respect to empirical evidence of renegotiation concession agreements. What is surprising is the fact, that all countries are EU members or were either a full fledges EU members or were negotiating to enter EU and therefore their level of institutional environment should be equal or similar and therefore the impact of institutional environment on renegotiation of concession agreements should have been similar.

3.1 Bucharest

Bucharest water concession agreement is a success story and as of 2006 it has 2372 km of main pipes and service pipes and around 670 km of pipelines related to more than 89,000 connections (Chiru, 2006, p. 123). The Apa Nova Bucuresti, a subsidiary (of Veolia Water) concession agreement was concluded in the year 2000 after a public bid. The Apa Nova Bucuresti, which was initially in 100% ownership of Bucharest municipality, was sold to private investors since it had big financial problems. The problems were due to poor infrastructure and the inability of the municipality to provide the necessary budgets for long-term investment and to repay the company's debts. In the process of selecting the concessionaire, Veolia Water's offer was approximately 30% lower than the next lowest bid submitted by International Water, and about 70% lower than the Suez Lyonnaise offer. The selected provider (Veolia Water) was selected based on a single variable, namely: the lowest realistic weighted average tariff valid for the entire duration of the concession. As of 2011, the service quality is above Romanian standards. Efficiency gains produced costs savings that amount of US$349 million and the concessionaire investment close to US$259 million. The service improved and the connection rate is high (93%), customer satisfaction rose from 50% to over 75%, and the operating costs are fourth lowest unit operation costs in Romania due to gains in labour productivity, which was pretty dramatic due to lay-offs during years 2000–2003. The typical household bill increased only modestly in the first 10 years of the concession (Earhardt, Rekas, & Tonizzo, 2011). In addition, the utility has the second slowest rate of unit energy cost increases, in 2007 collected 98% of bills, reduced waste and cut the level of nonrevenue water by nearly half. In the first nine years, it invested US$66 million in pipe replacement and other measures to prevent water leakage. Rough estimates were made to compare how water supply costs would be incurred if there was no privatization and as showed in table below, the costs of providing the same water supply services would have been US$349 million higher (Fig. 13.1).

Many attribute the success of Bucharest water concession to the provisions of the concession agreements and the team that was involved in the privatization process. These two variables fit nicely with the above presented empirical results for the renegotiation of concession agreements. The commitment of bureaucracy and the ability to work with outside investors are part of the variable that encompasses the quality of bureaucracy (Chiru, 2006). The provisions of the concession contract decrease the asymmetry of the information between the contracting parties and can be attributed to the existence of the regulator and also quality of the bureaucracy. IFC (the International Finance Corporation belongs to the World Bank) participated

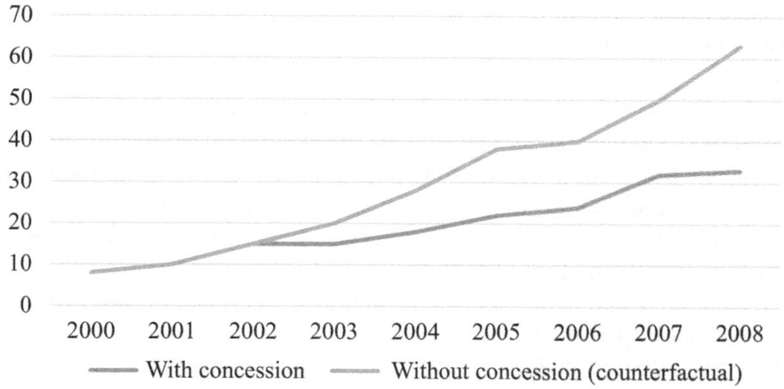

Fig. 13.1 Average monthly household bills with and without concession, 2000–2008 (Earhardt et al., 2011)

in the proposed privatization both in terms of the regulatory framework, transaction structure, price structure, and future liabilities in the provision of services. It also oversaw the supply procedure itself, which led the private investor into the sector (World Bank, 2002). The monitoring of the concession agreement and service level targets and goals is supervised by the Agency for the water and sewerage services in Bucharest. According to the World Bank publication Viewpoint (Earhardt et al., 2011), the senior officials in the municipality and in the utility were fully committed to a public—private partnership, and the municipality had advisors with strong reform credentials. What is more, mechanisms for regulation, monitoring and dispute resolution were also key to success and the neutral expert panel has been crucial in maintaining the relationship between the contracting parties (The World Bank, 2002). For example, the control of fulfilment of service levels is in the hands of ARBAC, the agency for the regulation of water supply and sewerage services, which does not have any control on tariff setting and its supervision of service level is disinterested and free of conflict of interest. There are many entities involved in tariff setting, but at the end, the International Experts Commission has the final say, which might contribute to expert tariff setting. The dispute resolution mechanism involves an economic regulator and a technical regulator housed in the municipal government, with recourse to an international panel of experts in case of appeal, which guarantees expertise and credibility in dispute resolution.

The concession agreement includes the following requirements:

- the Municipality of Bucharest remains the owner of water assets, and Apa Nova Bucuresti can use and benefit from the infrastructure, while at the same time it is obliged to take care of the management and the infrastructure renewal;
- concessionaire is required to achieve and maintain the performance indicators on European standards (levels of service) and invest in such a way to achieve the service levels, which are:
- potable water quality;

- pressure at the connection;
- continuity of service;
- coverage;
- time between notification of bursts/leak and reinstatement of water;
- water billed and metered as a percentage of water produced (Chiru, 2006).
- the methods of measurement, monitoring procedures, the compliance requirements, allowable exclusions and the remedial actions are set;
- the maintenance control ("supervision service levels") is carried out by a regulatory agency in Bucharest (ARBAC);
- tariffs are the result of various factors (national, local) but are supervised by the International Experts Commission;
- the contract is concluded for 25 years (i.e., until 2025), with the possibility of an extension of up to 37 years.

All in all, Bucharest water privatization is a success story. It required no renegotiations. One of the reasons seems to be that the international expert panel set the tariffs based on experience and due diligence which decreased the asymmetry of information among the private sector and the government, that the supervision of the service levels is in the hands of a regulatory agency, but has no jurisdiction on setting the tariffs, that goals are set in concession agreement and are therefore very transparent, that regulatory agency was in place at the time of negotiations, bureaucracy involved in the privatization was competent and committed to the privatization, there was also a loan form the World bank so the private financing was not exclusive and that there is no minimum income guarantee. Also, the dispute resolution mechanism gives credibility to the contract by ensuring that all the disputes are going to be adjudicated by international and disinterested experts. Although it seems that Romania is more corrupt that some of the countries described in the text below there is no evidence that corruption affected the water privatization process at the municipal level (Fig. 13.2).

3.2 *Grenoble*

Mayor of Grenoble signed a 25-year "affermage" agreement with COGESE, a subsidy of Lyonnaise des Eaux in 1989, without a bidding process. Soon after being awarded the privatization contract, COGESE increased the tariffs for 164% and then yearly increased tariffs slightly above inflation and in such a way that it could not justify the amount of investment (Lobina, 2006). In addition, later on in 1995 it was discovered by the Chambre Régionale des Comptes (regional audit body) that COGESE inflated costs and used fraudulent accounting practices (Lobina, 2006). The investment lagged behind and was substantially lower than in the years preceding privatization, and the amount spent by the private sector was only covering operational and maintenance costs. The contract was renegotiated in 1995 when corruption charges were made public and the new municipal elections came about.

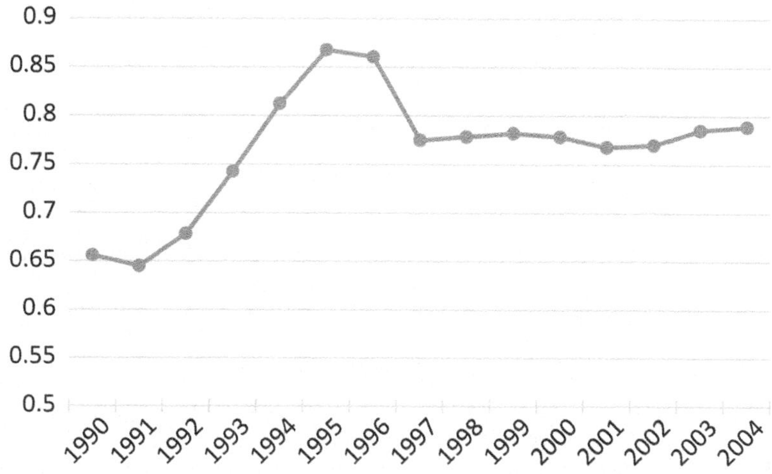

Fig. 13.2 Water supply tariffs, 1990–2004 (€ pcm, in real terms at 2004 value—taxes excluded, including fix amount for a consumption volume of 120 cubic meters) (Lobina, 2006)

Municipality did not cancel the "affermage" contract due to high contract cancelation fees. It formed a public—private partnership with existing supplier, but the situation did not improve for customers. Finally, in 2001, the water service was put in municipality hands once again after the highest instance court declared the concession agreement illegal and the concessionaire had to refund the customers for high fees (Lobina, 2006). Also, the mayor of Grenoble, Alain Carignon, was imprisoned and served 29 months of jail time being sentenced to 4 years due to corruption charges and the water tariffs were declared illegal by a French court (Bankwatch, 2019). After water distribution came into pubic hands once again, the tariffs, which were the highest in 1995 stayed constant (see Fig. 13.3) despite major innovation in the water distribution system as seen the Fig. 13.2.

The failure of Grenoble water privatization contract is due to many factors. One was corruption at the municipal level at the time when the concession was awarded without bidding and throughout the concession agreement duration, another was lack of transparency and lack of oversight, which led to undue tariff increases. What is more, the mechanism for changing tariffs was non-existent and tariffs were set by the municipality, which apparently benefited from the increased tariffs through corruption.[7] What makes matters even more puzzling is the fact that the water sources in Grenoble area are abundant and that the distribution system was not in a bad shape when affermage was awarded.

[7] In 1996, both the mayor of Grenoble at the time of the concession contract, Alain Carignon and Jean_Jacques Prompsy, then a Lyonnaise des Eaux, were convicted to go to prison on corruption charges by Cour de Cassation and spent some time in prison. Also, the Chambre Régionale des Comptes showed that contractual arrangements were design to secure concessionaire more than FF 1 billion excess profits in 25 years to the detriment of consumers and taxpayers (Lobina, 2006).

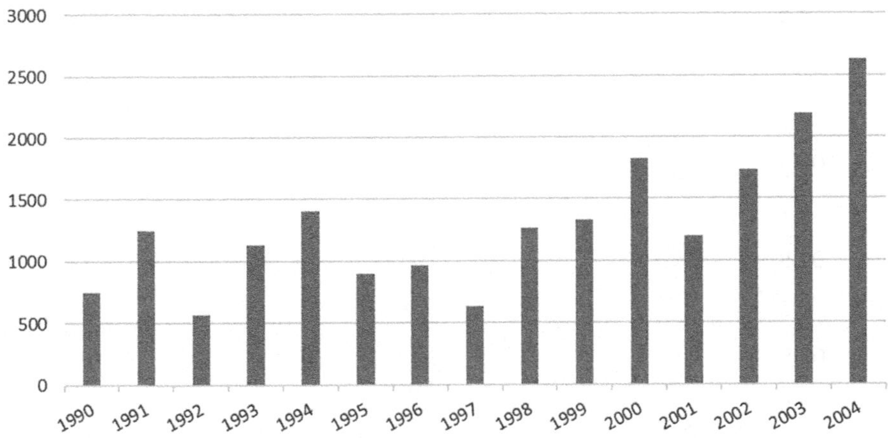

Fig. 13.3 Water supply pipelines renewed per year, 1990–2004 (linear meters) (Lobina, 2006)

3.3 Paris

Up to the nineteenth century water supply in Paris was in public hands, even though some private suppliers supplied water to the wealthy customers (Barraque, 2012). Nineteenth-century Paris had a public water service which was chiefly used to 'wash the city clean', fight fires, water the parks and feed public fountains. There were also private services providing water to wealthy residents but a universal domestic water supply still did not exist.

There are 2 million people living in Paris, with 3.5 million daily commuters and tourists and 93,000 subscribers to water, daily consumption is around 550,000 m^3 and 1500 km long network is built with 470 km of aqueducts (Le Strat, 2017). In 1984, water supply in Paris was awarded to two private companies, the area to the left of the Seine to Lyonnaise des Eaux, the larger area on the right side of the Seine was awarded to Compagnie Generale. The leases were awarded without a bid and for 25-year period. Private companies were managing the distribution and infrastructure, on the other hand, was financed with public funds. However, construction of infrastructure was arranged by the water companies themselves (Schiffler, 2015). Despite water bill doubling in 25 years, the private sector replaced old lead pipes, and all of 93,000 m in Paris were replaced, the leakages were down 3% and water quality increased.

Water leases expired in 2010 and they were not renewed. Public utility Eau de Paris is now supplying water in Paris. The municipalization brought about 15% savings the first year (approximately 76 million per year), and water tariffs were lowered by 8% in 2011 (Bankwatch, 2019), which is mainly due to the fact that the pipelines were upgraded in the last 25 years and now need less investment, and due to the mandatory bidding process for any kind of public investment (Barraque,

2012). The positive outcomes or results of the municipalization and changed regulation show in easier control and oversight of the water supply, revenue must be reinvested, performance indicators are introduced, together with benchmarking to France and the rest of EU and annual activity report is introduced.

To sum up, it seems that corruption was a big part of water concession agreements' bidding and tariff setting in Grenoble and Paris. This was partly solved in 1993, when the Socialist party came to power in France. They enacted the law on prevention of corruption and promotion of transparency in public life. The law dictated municipalities to award concessions by bidding process, limited the duration and extension of the privatization contracts and limited political contributions. The laws that were passed in order to avoid the situation that happened in Grenoble and Paris. In addition, a law passed in 1995 compelled water utilities to submit operational data to municipalities. Finally, in 2006, ONEMA, National Water Agency, which regulated the water sector, was created (Schiffler, 2015).

3.4 Portugal

Portugal is one of the few European countries where an empirical analysis was performed on the dataset of 31 concession agreements during years 1995–2005 (Andrade et al., 2018). Some concession agreements were for water services only, some for sanitation and some for both services together. In Portugal, municipalities are responsible for water supply at the retail level. Most of the water is supplied by the municipalities directly, but there is a growing number of concession agreements. Portugal has registered a significant growth in the coverage of water and waste water networks and during the 1990s up to 2010. The coverage of water services increased from 82% to 97%, and in case of the waste water, from 58% to 69%. Andrade et al. (2018) analyse 41 incidences of renegotiations of water and wastewater concession agreements.[8] The first renegotiation of the concession was in year 2000, 5 years after the first concession was awarded. The frequency of renegotiations peaked in 2012. The analysis confirms that larger and longer concession agreements are more likely to be renegotiated, and that election period increases the likelihood of renegotiations. Furthermore, they confirmed that better institutional and legal environment decreases the probability of renegotiations and that during the financial crisis or from 2011 on, the renegotiations increased. However, two variables were most notable in renegotiations, the demand variation and unforeseen investments, which usually result in tariff increase and direct financial compensa-

[8] The chapter considers renegotiations to be a change in an initial contract, but not a tariff increase, for example which is linked to automated inflation based formula and similar.

tion of private sector by the government or the extension of the contract.[9] The chapter concludes that with respect that due to incompleteness of the concession contracts renegotiations are inevitable rules should be put in place regarding what variables are even up to renegotiation and a mechanism for renegotiations, especially when there is variation in demand and unforeseen investment, should be put in place. In addition, the asymmetry of information should be decreased, by motivating the private sector to provide more information before conclusion of the concession contract and during the concession agreement. The final recommendation of Andrade et al. (2018) is that expectations of contracting parties at the onset of the concession agreement negotiations should be realistic and not too optimistic, which could be amended by independent due diligence supervised by regulator.

4 Discussion with Conclusion

Even though some concession agreements seem to be efficient, expanding the water network, providing high quality water at reasonable costs/tariffs, they are prone to renegotiations. Renegotiations are not bad per se since they could be Pareto improving, but on the other hand, they could be a consequence of rent seeking and the end result of renegotiations could decrease the welfare in the society. Finding out what prevents renegotiations of concession agreements is therefore of importance, especially to prevent rent-shifting motivated renegotiations.

Even though theoretical prediction points out that in developed institutional environment, the renegotiation of concession agreements should be low and both, the government and private sector have low motivation for renegotiations, for example in EU, where all member countries have to abide to minimal institutional standards to even join the EU, the anecdotal evidence seems to points in a different direction. For example, Paris and Grenoble privatization contracts were renegotiated many times and the consequences were lower than agreed on investments and increase tariffs in the water sector. However, even though the rule of law in France at that time was high, it seems that the local corruption was prevalent, since some of the main actors even landed in jail for some time. Also, there were some other variables pointing to weak institutional environment, for example lack of bidding process, non-existence of regulatory agencies and non-transparent tariff setting. We can conclude that the strength of the institutional environment, when predicting success of the concession agreements should be determined at the local, and not at the national level, and no concession agreement process should be started without proper institutional setting at the local level. Budapest concession agreement also points to the importance of local conditions. As the literature points out, the water concession in

[9] Unfortunately, Andrade et al. (2018) did not provide the data which renegotiations were initiated by the government and which by the private sector.

Bucharest was well prepared at the outset, with bureaucracy fully committed to the privatization process, with existing regulatory agency and with mechanism in which tariffs are supervised by outside experts. However, in Bucharest, beside institutional environment, the provisions in the concession contract seem also to contribute to its relative success. For example, the concession agreement sets the level of service that the concessionaire has to achieve and their success is supervised by the regulatory agency, which has no power to set the tariffs, which are set by outside experts and the dispute resolution is clearly defined with involvement of outside experts on the appeal level. In Portugal, empirical evidence shows that better institutional and legal environment decreases the probability of renegotiations However, two variables were most notable in renegotiations, the demand variation and unforeseen investments, which usually result in tariff increase and direct financial compensation of private sector by the government or the extension of the contract, which is not so surprising in view of Guasch et al. (2003) and Guasch and Straub (2006) since the demand variation and unforeseen investments are more or less captured in their variable that measures investment and variable that measures economic shocks. Even though it should be pointed out, as authors pointed out (Andrade et al., 2018) that pre-concession-award due diligence is extremely important in order to deflate expectations of the government and private sector and therefore negative consequences of demand shocks and unforeseen demand.

It does seem that all four presented cases to a certain degree conform to existing empirical evidence on the renegotiation of the concession agreements. The development of institutional environment matters, especially at the local level, the quality of regulation and supervision of the concessionaire and involvement of independent experts also and so do some other provisions in concession agreement, for example level of investment, regulation of profits and minimal income guarantee. What might not be captured in the empirical evidence presented above are some other different provisions of concession agreements and their effect on renegotiations as Bucharest example might indicate. The same goes for the influence of local level of corruption and/or quality of bureaucracy since concession agreements are mostly awarded by municipalities and not the government and therefore the variables that measure institutional environment should be locally measured. Furthermore, it is not always clear from the four cases that were presented, whether the renegotiation was initiated either by private or government sector, especially in Portugal case, which opens the door for future research. Recommendation should certainly be that local institutional environment should be examined and improved before starting the process to award concessions and that content of concession agreements should be modeled after good practices in this field.

Acknowledgements I would like to thank Zoja Bajželj and Maksimilijan Gale for excellent research assistance. All errors are of course my own.

References

Alchian, A. A. (1965). Some economics of property rights. *Il Politico, 30*(4), 816–829.

Alchian, A. A., & Demsetz, H. (1972). Production, information costs and economic organization. *American Economic Review, 62*(5), 777–795.

Andrade, I., Cruz, C. O., & Saramento, J. M. (2018). Renegotiations of water concessions: Empirical analysis of main determinants. *Journal of Water Resources Planning and Management, 144*(11). https://doi.org/10.1061/(ASCE)WR.1943-5452.0000999

Bankwatch. (2019). *French Water Concession.* Retrieved July 23, 2019, from https://bankwatch.org/public-private-partnerships/case-studies/french-water-concessions

Barraque, B. (2012). Return of drinking water supply in Paris to public control. *Water Policy, 14*(6), 903–914.

Chiru, E. (2006). Water supply of Bucharest - Past, present, future: A study case. In P. Hlavinek, T. Kukharchyk, J. Marsalek, & I. Mahrikova (Eds.), *Integrated urban water resources management* (pp. 119–130). Dordrecht: Springer.

Coase, R. H. (1937). The nature of the firm. *Economica, 4*(16), 386–405.

Cochabamba water privatization. (2019). Retrieved July 22, 2019, from https://mru.org/courses/development-economics/cochabamba-water-privatization

Earhardt, D., Rekas, M., & Tonizzo, M. (2011). *Water in Bucharest* (Viewpoint). The World Bank Group. Accessed July 23, 2019, from https://www.google.com/url?sa=t&rct=j&q=&esrc=s&source=web&cd=1&ved=2ahUKEwiisJCFscrjAhUCJVAKHQKoDfUQFjAAegQIARAC&url=https%3A%2F%2Flibrary.pppknowledgelab.org%2Fdocuments%2F2035%2Fdownload&usg=AOvVaw2gWcNoelx4iW7BoC3xRSjy

Foreman-Peck, J., & Millward, R. (1994). *Public and private ownership of British industry 1820-1990.* Oxford: Clarendon Press.

Getzner, M., Köhler, B., Krisch, A., & Plank, L. (2018). *Comparison of European water supply and sanitation systems. Final report (abridged version). Information on environmental policy, 197b.* Vienna: Vienna Chamber of Labor.

Grossman, S. J., & Hart, O. D. (1986). The costs and benefits of ownership: A theory of vertical and lateral integration. *Journal of Political Economy, 94*(4), 691–719.

Guasch, J. L. (2004). *Granting and renegotiating infrastructure concessions: Doing it right.* Washington: WBI Development Studies.

Guasch, J. L., Laffont, J.-J., & Straub, S. (2007). Concessions of infrastructure in Latin America: Government-led renegotiation. *Journal of Applied Econometrics, 22*(7), 1267–1294.

Guasch, J. L., & Straub, S. (2006). Renegotiation of infrastructure concessions: An overview. *Annals of Public and Cooperative Economics, 77*(4), 479–493.

Guasch, J. L., Straub, S., & Laffont, J.-J. (2003). Renegotiation of concession contracts in Latin America. Policy research working papers . Washington, DC: World Bank Publications. https://doi.org/10.1596/1813-9450-3011

Hart, O., & Moore, J. (1990). Property rights and the nature of the firm. *Journal of Political Economy, 98*(6), 1119–1159.

Kishimoto, S., Lobina, E., & Petitjean, O. (2015). *Our public water future: the global experience with remunicipalisation. Services International Research Unit (PSIRU), Multinationals Observatory, Municipal Services Project, European Federation of Public Service Unions (EPSU).* Amsterdam, London, Paris, Cape Town, Brussels. Retrieved July 22, 2019, from https://www.google.com/url?sa=t&rct=j&q=&esrc=s&source=web&cd=1&ved=2ahUKEwjCo97ig8jjAhWIL1AKHZTqCl0QFjAAegQIAhAC&url=https%3A%2F%2Fwww.ft.dk%2Fsamling%2F20151%2Falmdel%2Fefk%2Fnilag%2F408%2F1667841.pdf&usg=AOvVaw0TS6nwXhmRtuShG7lp6rf6

Klein, B., Crawford, R. G., & Alchian, A. A. (1978). Vertical integration, appropriable rents, and the competitive contracting process. *Journal of Law and Economics, 21*(2), 297–326.

Laffont, J. (2003). Enforcement, regulation and development. *Journal of African Economies, 12*(2), ii193–ii211.

Le Strat, A. (2017). *The remunicipalisation of Paris' water supply service*. Paris: Mairie de Paris. Retrieved July 23, 2019, from https://research.ncl.ac.uk/media/sites/researchwebsites/gobacit/Anne%20Le%20Strat.pdf

Lobina, E. (2006). D21: Watertime case study – Grenoble, France. *Watertime*. Retrieved July 23, 2019, from https://www.google.com/url?sa=t&rct=j&q=&esrc=s&source=web&cd=1&ved=2ahUKEwjVkcXHv8rjAhUOEVAKHX6DDdQQFjAAegQIARAC&url=http%3A%2F%2Fwww.watertime.net%2Fdocs%2FWP2%2FD21_Grenoble.doc&usg=AOvVaw20Yz-iLUdFUcRjzJGxLesY

Lobina, E. (2017). Water remunicipalisation: between pendulum swings and paradigm advocacy. In S. Bell, A. Allen, P. Hofmann, & T.-H. The (Eds.), *Urban water trajectories* (Future city) (pp. 149–161). Cham: Springer.

Nickson, A., & Vargas, C. (2002). The limitations of water regulation: The failure of the Cochabamba concession in Bolivia. *Bulletin of Latin American Research, 21*(1), 99–120.

Schiffler, M. (2015). *Water, politics and money: A reality check on privatization*. Frankfurt: Springer.

The World Bank. (2002). *Implementation Completion Report (TF-20515; SCL-40790) on a Loan in the Amount of US$ 25.0 Million to Romania for a Bucharest Water Supply Project* (Report No. 23912). Retrieved July 23, 2019, from http://documents.worldbank.org/curated/en/486191468293464874/pdf/multi0page.pdf

Williamson, O. E. (1987). Transaction cost economics: The comparative contracting perspective. *Journal of Economic Behavior & Organization, Elsevier, 8*(4), 617–625.

Zajc, K. (1997). *Private sector participation options in the water sector in transition economies*. Doctoral dissertation, George Mason University, Fairfax, VA.

Zajc, K. (2004). *Privatization and water infrastructure*. Trieste: EST – ZTT, Editoriale Stampa Triestina.

Zerah, M. H., & Graham-Harrison, K. (2001). The Buenos Aires concession: Private sector serving the poor (Case Study). Retrieved July 22, 2019, from https://www.wsp.org/sites/wsp/files/publications/sa_buenos.pdf

Index

K. Eman et al. (eds.), *Water, Governance, and Crime Issues*,
https://doi.org/10.1007/978-3-030-44798-4

W

The manufacturer's authorised representative in the EU is Springer Nature Customer Service Centre GmbH, Europaplatz 3, 69115 Heidelberg, Germany. If you have any concerns regarding our products, please contact ProductSafety@springernature.com

Printed and bound by CPI Group (UK) Ltd, Croydon, CR0 4YY
30/09/2026
02225450-0017